Disaster Risk and Impact Management

Approaches, Tools and Strategies

THE EDITORS

Dr. Anil Kumar Gupta, Head - Division of Policy Planning, National Institute of Disaster Management, New Delhi, India (Former Reader & Head, Director, Institute of Environment & Development Studies, Bundelkhand University, Uttar Pradesh, India).

Dr. Vinod K. Sharma is Senior Professor, Disaster Management at Indian Institute of Public Administration, New Delhi and Executive Vice Chairman, Sikkim State Disaster Management Authority and Vice Chancellor of Devi Ahilya University Indore.

Sreeja S. Nair, Consortium Coordinator, Myanmar Consortium for Disaster Risk Reduction Capacity Building, Yangon, Myanmar (Assistant Professor, National Institute of Disaster Management, New Delhi).

Disaster Risk and Impact Management

Approaches, Tools and Strategies

Editors
Anil K. Gupta
Vinod K. Sharma
Sreeja S. Nair

2018
Daya Publishing House®
A Division of
Astral International Pvt. Ltd.
New Delhi – 110 002

ISBN: 9789351240013 (International Edition)

Publisher's note:

Published by   : **Daya Publishing House®**
                        A Division of
                        Astral International Pvt. Ltd.
                        – ISO 9001:2015 Certified Company –
                        4736/23, Ansari Road, Darya Ganj
                        New Delhi-110 002
                        Ph. 011-43549197, 23278134
                        E-mail: info@astralint.com
                        Website: www.astralint.com

Laser Typesetting : **Classic Computer Services**, Delhi - 110 035

Praface

In India, the devastation that could be caused by industrial disaster became poignantly evident during the Bhopal Gas Tragedy in 1984. Following Bhopal, laws and regulations have been made more stringent, but the state of industrial safety and management measures on the ground are such that chances of major and minor accidents, in almost every corner of the country are considerable. And it would be communities near the industries which would probably bear the brunt of the impacts of such accidents. Community preparation is a significant, but often overlooked or underemphasized input for reducing disaster impacts.

Disaster is any occurrence that causes damage, ecological disruption, loss of human life and deterioration of health and health services on a scale sufficient to warrant an extraordinary response from outside the affected community. A disaster as an occurrence such as hurricane, tornado, storm, flood, high water, wind driven water, tidal wave, earthquake, drought, blizzard, pestilence, famine, fire, explosion, volcanic eruption, building collapse, transportation wreck, or other situation that causes human suffering or creates human needs that the victims cannot alleviate without assistance.

The present book "Disaster Risk and Impact Management: Approaches, Tools and Strategies" comprises of three sections: Disaster Risks: Overview, Case Studies and Strategic Approaches. Experts in the relevant fields have given critical appraisal of the various aspects of disasters risk and its management. The scientists, teachers, scholars, administrators and policy makers dealing with disaster risk and impact management find this book very useful and informative.

I am thankful to all the chapter contributors and co-editor for all support and motivation. cooperation and support of editorial team of Astral International Pvt. Ltd., New Delhi and especially Mr. Anil Mittal and Mr. Prateek Mittal deserve special thanks.

Dr. Anil K. Gupta

Principal Editor

Contents

Strategic Approaches

List of Contributors

1. **Agwata, Jones F. (Dr.)**

   Kenyatta University, Department of Environmental Sciences, PO BOX 43844-00100 GPO, Nairobi, Kenya

   E-mail: agwatas@yahoo.co.uk

2. **Birol, Ekin**

   Department of Land Economy and Homerton College, University of Cambridge, Hill Road, Cambridge CB2 8PH, UK

   E-mail: eb337@cam.ac.uk

   Tel.: +44 1223 507230; *Fax:* + 44 1223 507206

3. **Breguet, A.**

   The University of Lausanne/Institute for Geomatics and Risk Analysis, Switzerland (UNIL-IGAR)

4. **Chaudhary, B.S. (Dr.)**

   Department of Geophysics, Kurukshetra University, Kurukshetra, Haryana, India

   E-mail: bsgeokuk@yahoo.com

5. **Cheema, M.A.**

   The World Conservation Union-Pakistan (IUCN-P)

6. **Dar, Shabir Ahmad**

   National Research Centre for Agroforestry, Gwalior Road, Jhansi – 284 003, India

7. **Dubey, Kumud (Dr.)**

   Centre for Social Forestry and Eco-Rehabilitation, 337-Ashok Nagar, Allahabad – 211 001, India

   E-mail: dkumud@yahoo.com

8. **Dubois, J.**

   The University of Lausanne/Institute for Geomatics and Risk Analysis, Switzerland (UNIL-IGAR)

9. **Guleria, Sushma**

   National Institute of Disaster Management, IIPA Campus, New Delhi, India

   E-mail: sushguleria@gmail.com

10. **Gupta, Anil K. (Dr.)**

    National Institute of Disaster Management, IIPA Campus, New Delhi, India

    envirosafe2007@gmail.com

    Tel.: +91-011-23724311, +91- 9868207006

11. **Jaboyedoff, M. J.**

    The University of Lausanne/Institute for Geomatics and Risk Analysis, Switzerland (UNIL-IGAR)

12. **Jaubert, R.**

    The Graduate Institute for Development Studies, Geneva Switzerland (IUED)

13. **Kathil, Nidhi**

    J. C. Bose Institute of Life Sciences, Department of Biochemistry, Bundelkhand University, Jhansi – 284 128, India

14. **Kaur, Ravneet (Prof.)**

    Maulana Azad Medical College, New Delhi – 110 002, India

15. **Kishore, Jugal (Prof. Dr.)**

    Maulana Azad Medical College, New Delhi – 110 002, India

    E-mail: drjugalkishore@gmail.com

16. **Koundouri, Phoebe**

    Department of International and European Economic Studies, Athens University of Economics and Business, 76, Patission Street, Athens 104 34, Greece

    E-mail: pkoundouri@aueb.gr

    Tel.: +30 210 8203455; *Fax:* +30 210 8214122

17. **Kountouris, Yiannis**

   Department of International and European Economic Studies, Athens University of Economics and Business, 76, Patission Street, Athens 104 34, Greece

   E-mail: ykountouris@aueb.gr

   Fax: +30 210 8214122

18. **Lodhia, Shital (Dr.)**

   Centre for Development Alternatives, Ahmedabad, India

   E-mail: s_lodhia@rediffmail.com

19. **Nair, Sreeja S.**

   National Institute of Disaster Management, IIPA Campus, New Delhi, India

   E-mail: sreejanair22@gmail.com

   Tel.: +91-9810079551

20. **Newaj, Ram (Dr.)**

   National Research Centre for Agroforestry, Gwalior Road, Jhansi – 284 003, India

   E-mail: ramnewaj@nrcaf.ernet.in, ramnewaj@gmail.com

21. **Pandey, J. S. (Dr.)**

   National Environmental Engineering Research Institute, Napgur, India

22. **Peduzzi, P.**

   U.N. Environment Programme -GRID/EUROPE, Division of Early Warning and Assessment

23. **Qureshi, R. A.**

   The World Conservation Union-Pakistan (IUCN-P)

24. **Sehgal, Vinay, K. (Prof. Dr.)**

   Division of Agricultural Physics, Indian Agriculture Research Institute, Pusa, New Delhi, India

25. **Singh, V.K.**

   Centre for Social Forestry and Eco-Rehabilitation, 337-Ashok Nagar, Allahabad – 211 001, India

26. **Singh, Brijesh K.**

   J. C. Bose Institute of Life Sciences, Department of Biotechnology, Bundelkhand University, Jhansi – 284 128, India

27. **Singh, Shweta (Dr.)**

School of Social Work, Loyola University Chicago, IL 60611

E-mail: ssingh9@luc.edu

Tel.: +312-915-7645

28. **Srivastav, Anubha**

Centre for Social Forestry and Eco-Rehabilitation, 337-Ashok Nagar, Allahabad – 211 001, India

29. **Sudmeier-Rieux, K. (Dr.)**

The University of Lausanne/Institute for Geomatics and Risk Analysis, Switzerland (UNIL-IGAR)

30. **Tyagi, Pallavee (Dr.)**

Garg Institute of Engineering and Technology, Ghaziabad, U.P., India

E-mail: palavee1@rediffmail.com

31. **Verma, Manish**

Department of Natural Resource Management; Institute of Environment and Development Studies, Bundelkhand University, Jhansi – 284 128, India

E-mail: ndmk143@yahoo.com, taurian_bravo@yahoo.co.uk

32. **Yunus, M. (Prof. Dr.)**

School for Environmental Sciences, BB Ambdekar (Central) University, Lucknow, India

Disaster Risks: Overview

1

Disaster Risk Management for Sustainability

Anil K. Gupta, Sreeja S. Nair and Vinod K. Sharma

Concept of Environmental Disaster Reduction, natural and man-made, is now suggested to includes the identification and characterisation of potential causes of hazards or risk, maximum probable environmental impact, consequence analysis, prevention and control systems, impact minimisation planning, emergency preparedness and post-disaster management.

The lacunae here is that, these techniques for assessing the risk and potential of natural disaster, are highly expensive and give little emphasis on the cause, control or hazard prevention. Ecological risk assessments traditionally have been used to evaluate the adverse impacts of new chemicals or old contaminants on a particular environment. But a growing realisation that the degradation of eco-systems, and the onset of many natural disaster can be caused not only due to chemical factors but also due to physical and biological factors, has created a demand for the disaster reduction strategies based on ecological risk assessment. In addition, the existing focus emphases mainly the importance of relief-after-the-event strategy and rehabilitation. Separative consideration of of natural and man-made hazards in disaster management policy and non-consideration of disaster risks in industrial or developmental planning makes the problem complex and increases the vulnerability.

Assessment of hazards, especially the ones based in nature has been a concern for the scientific community for a long time. Better instrumentation, global networks, collaboration among the relevant institutions and agencies over the decades have resulted in improved hazard assessment techniques and data. Modern technology offers increasingly simple, reliable and low-cost methods for disaster mitigation. Yet paradoxically, the human and environmental impact of such events continues to increase at an alarming rate in almost all countries. There is still more to do in knowing the hazards, especially in relation to slow-onset disasters or environmental-natural hazards such as drought, flood, fire and ecological degradation (Aysan, 1993). However, understanding the vulnerability and damage-risk is only part of the picture. The decade's aim, that is, the *'reduction of natural disasters'* can not be achieved unless there is good an understanding of what are the interactive influences and their root causes leading to the prevalence of natural hazard.

The distinction between natural hazards or disasters and their man-made (or technological) counterparts is often difficult to sustain. Many researches have called the 'naturalness" of natural disasters into questing, though the circumstances of purely anthropogenic hazards, such as oil spills and chemical explosions, are usually very different from those of, for example, earthquake and floods. Nevertheless, in terms of the consequences there is a sizeable overlap (Alexander, 1993). Hence, much that is of benefit to the study of natural catastrophe can be learned from technological risks and disasters, and from the manner these interact with their natural counterparts. Man-made risks are important here as the technology has created new sources of risk, and increased old ones. It is also considerable that many aspects of disaster management and risk reduction, developed in response to technological impacts, are applicable to natural disasters.

Understanding Natural Hazards and Environmental Consequences

Natural hazards are those events that occur in nature and are capable of producing injury or death to people and/or damage to property. Human activities and ecological deterioration are likely to alter the frequency, increase or decrease their severity or intensity, alter the damage distance, influence the rate of exposure and vulnerability of hazard exposed persons and property (Petak and Atkisson, 1987). At the simplest level, humanity has the options either to live in harmony with the natural environment, by a symbiosis in which ecological life support systems are enhanced, or to exploit resources by a form of parasitism, in which hazards are ignored until they strike (Alexander, 1993). The former provides opportunities for sustainable development and protection against environmental extremes (natural disasters), while the later relies on careless exploitation of non-renewable resources, places little emphasis on hazard prevention and more on post-disaster relief, rescue and rehabilitation

Many small to moderate earthquakes have been provoked by the superpowers' programmes on nuclear testing, and hence, the size of the interaction at seismic network also reflects investment in intelligence work designed to monitor such trials. Some induced seismicity is, however, inadvertent (Judd, 1974). The phenomenon was first noticed in Greece in 1929, and on the filling of the Hoover Dam on the Colorodo River, which provoked about 600 earthquakes, the largest of which reached a magnitude of 5.0. In 1967 in West Central India, the Koyna Reservoir, which had a capacity of 2,780 million meter cube of water, caused a magnitude 6.5 earthquake which killed 177 people, injured 2,200 and left thousands homeless (Gupta and Combs, 1976). Reservoir induced seismicity can be a problem when hydraulic conductivity is strong are the way down to highly fractured rocks situated deep beneath the impoundment, and where water pressure is high in the saturated clefts between rock blocks. Whether caused by the seasonal increase of inland rivers or the storm rise of coastal waters, flooding results from the inability of these soil, vegetation, or atmosphere to absorb the excess water (Cornell, 1982). Adjustment to floods can broadly be classified into structural and non-structural, according to whether they use engineering or administrative methods (Thampapillai and Musgrave, 1985). Long-terms measures of flood prevention are based on the management of catchment involving: (1) afforestation, (2) other vegetational changes (3) agricultural measures (4) management of snow-covered areas, (5) management of urban areas, and (6) implementation of efficient water resource conservation strategy. Land use control combined with integrated water management programme, can be one of the most important activities in reducing flood hazards. Statutes, regulations and public motivations campaign can be employed in this line. Clearly, flood mitigation involves some difficult dilemmas. For instance using dams as a structural measures of flood control is problematic. It is difficult to decide between the paralle absence of a large number of head-water dams, which may exert a limited effect on a trunk river, or a measure down-stream dam, which may be vulnerable to failure and collapse. In general, up-stream dams are best at controlling erosion and sedimentation, while down-stream ones act best to restrict flooding. The later may need to be kept half empty if they are to reduce high discharge rates, but this way this may limit their value for water supply, power generation or recreational uses. Perhaps for this reason, only 17 of India's 1,554 large dams were built with objectives of flood control.

Drought can be defined as a condition of abnormal dry weather resulting in a serious hydrological imbalance, with consequences such as losses of standing crop and shortage of water needed by people and livestock (Alexander, 1993). In hydrological terms, a precipitation drought is caused by lack or rainfall, a run-off drought by lack of stream-flow and an aquifer drought by lack of ground water. Drought result from a prolonged period of moisture deficiency. One concept is of that particular use in the identification of drought condition is albedo. At the global scale albido represents the total power of the earth's surface and its atmosphere to term back incoming solar radiation. As moisture conditions can often be detected by remote sensing as an increase in surface reflectivity, caused by desiccation of the land surface and wilting or death of vegetation and crops (Idso *et al.*, 1975).

Main draw back of the existing drought management practices is the non-consideration of long term key influences causing droughts and the expansion of drought affected areas. Scientifically, it can be considered that floods are also one of the principle causes of drought. The existing systems involved only post-event relief or pre-disaster planning for emergencies, but no proper emphasis has been given on the strategies of water-resource conservation practices to prevent these hazards (Gupta *et al.*, 1997). Many episodes of landslides and most of the forest fires are the environmental consequences of human activities.

Relationship between Natural and Man-made Hazards

Natural Hazards are known to be as old as nature but man-made hazards are of recent origin. Beginning with the industrial era man's activities started to cause considerable changes in the natural environment. Presently they have reached such a high level that their physical, chemical or biological effects become sometimes similar to the effects of natural hazards and even to the natural phenomena causing disasters (Jovanovic, 1988).

Table 1.1: Interacting influences on consequences between the natural and man-made hazards.

Natural Hazards	Man-Made Hazards
Tsunami	☆ Coral reef destruction
	☆ Destruction of beach forest and other protective features
	☆ Low settlement location
☆ Natural forest fire	☆ Bad planning of long distance transport electric lines
☆ Fires due to lightning	☆ Poor management of forest areas
☆ Fire due to volcanic explosions	☆ Lack of protective corridors
	☆ Negligence of fire sensitive areas
☆ Landslide	☆ Bad road engineering
☆ Rock slides	☆ Open mining
	☆ Deforestation
	☆ Poor drainage system
Earthquakes	☆ Underground nuclear testing
	☆ Bad planning of water development projects
	☆ Deep mining
	☆ Sensitive material and construction design of buildings
	☆ Dense settlement
Floods	☆ Deforestation
	☆ Expansion of denuded areas
	☆ Destruction of vegetation cover
	☆ Erosion of soil cover
	☆ Environmental modifications
	☆ Poor design of community infrastructure
Droughts	☆ Destruction of vegetation
	☆ Environmental modifications
	☆ Over exploitation of water resource
	☆ Pollution/siltation oriented eutrophication leading to death of water body
	☆ Alteration of evapotraspiration system
☆ Cyclones	Similar to hazards described under earthquakes, floods, tsunamis
☆ Hurricanes	
☆ Typhoons	
☆ Tornadoes	

Contd...

Table 1.1–*Contd...*

Natural Hazards	Man-Made Hazards
Modifications of temperature "homeostasis" of planet earth	☆ Technologic trace gases and biomass burning influences on formation of "greenhouse" effect and climate change
Ozone layer variations	☆ Technological influence on trace gases concentration leading to ozone layer depletion and UV-B increase and biological perturbations
Soil carbon depletion	☆ Breaking of biogeochemical carbon cycle in agricultural areas
	☆ Over removal of biomass resources and man-made fires

Table 1.1 reveals the interacting mixed influence on onset or severity of consequences between the natural and man-made hazards. Interactions between these two types of disasters, based on the effects are presented in Table-1.2. Aspects of these relationships are presented (Modified after Jovanovic, 1988) in Table-1.3.

Table 1.2: Classified relationship between man-made and natural hazards according to effect.

Man-made Activities		Natural Phenomena
Changes of location of	Changes caused in human	☆ Earthquake
☆ Development practice	activities-behaviour	☆ Droughts
☆ Development projects	or history due to	☆ Volcanoes
		☆ Floods
		☆ Tsunamis
Changes in civilisations locations patterns, ethnical invasions, migrations, wars, etc.	–//– ⇓	Climate belts shifts of geosolar origin
Challenges from natural environment: Water, food, fodder, energy supply, environmental quality difficulties	Severity of consequences increased by	**Challenges from man-made environment:** ☆ Epidemics ☆ Fires ☆ Pollution

Table 1.3: Aspects of relationship between man-made and natural hazards.

Man-made Hazards		Natural Hazards
Water pollution Water rehabilitation	*Reversible*	☆ Sewage disposal and river biota destruction
		☆ Water treatment
Air pollution	*Irreversible*	Acid rain

Arithmetic effect
Man-made event + natural event = Disaster

Resonance Effect
Small explosion-resonance-earthquakeLow magnitude-resonance-high magnitude

Disaster Assessment and Environmental Risk Mapping Based Guidelines

Appropriate evaluation of a hazard requires determination of the occurrence of a natural event at its various intensity levels. To appropriately assess the levels of risk associated with natural hazard exposures in India, it is required to carry out following studies :

1. Identification and characterisation of natural hazard problems,

2. Selection of potential hazards and past accident analysis,

3. Study and formulation of cause-effect interactions and influence network modeling,

4. Exposure-response assessment and vulnerability studies,

5. Identification and measurement of the major primary, secondary and high order effects,

6. Identification of the cost and characteristics of the appropriate technologies for preventing/mitigating the hazards,

7. Assessment of associated public policies and improvements based on ecological/scientific and anthropological factors.

When it is considered that both type of disasters, natural and man-made, are of environmental origin, and have inter-influencial relationship, it becomes imperative to adopt an integrated approach for hazard reduction/prevention and hazard control (from becoming a disaster). This has been the experience that unplanned development and other human activities have increased the occurrence and intensity of disasters to an alarming consideration. A hazard becomes a disaster either due to increases vulnerability of land uses. "Environmental Risk Mapping Based Developmental Planning" is a preventive approach for designing Disaster Management based sustainable development.

In India, 'Ecological Carrying Capacity based Developmental Planning' concept has been initiated by a CSIR Institute - NEERI, and prepared plans for Doon Valley, National Capital Region (NCR), Jamshedpur region and Satna region etc. In an another effort to plan better industrial siting based on environmental considerations, Central Pollution Control Board designed a tool 'Zoning Atlas for Siting of Industries' and prepared the Zoning Atlases for around 20 districts in the country. However, major thrust of both these activities was on supportive capacities and positive effects of development. 'Zoning Atlas' procedures consider the existing land-uses for industrial siting but fail to evaluate short-term and long-term *vice-versa* impacts giving rise to potential hazards. Disaster risk due to siting of hazardous/chemical installations and natural hazards (increase in risk and vulnerability due to man caused stresses) potential of a site is required to be considered imperatively. Ecological "Risk Assessments may be the tool for understanding the risk of slow-onset disasters of environmental origin. 'Carrying Capacity based Planning' is a very expensive and time taking project; and 'Zoning Atlas' is a district planning tool. But as the inter-district or inter-regional influences are now very strong and citizen settlement or other developmental activities add to it, 'Zoning Atlas' fails to project any prediction on the future land-uses and hazard probabilities. Therefore a new tool "Environmental Risk Mapping Based Guidelines for Developmental Planning" is conceptualised. The procedural framework for this is designed by the authors. Important among the aspects to be considered in 'Risk Mapping' are as :

☆ Resource (natural, human and infrastructural) supportive capacities,

☆ Assimilative capacities (resilience, resistance and dilution/dispersion),

☆ Risk zoning of hazardous installations/highly polluted sites (damage/ impact distance),

☆ Risk intensity of natural and ecological hazards,

☆ Population/land-uses vulnerability and hotspots (categorisation),

☆ Land-use and land-cover classes, Waste lands,

☆ Tribal settlements, Mining sites, Hazardous waste sites *etc.*,

☆ Historical and cultural monuments, Dense settlements,

☆ Watershed, Airshed, Meteorological characteristics,

☆ Area specific notifications/regulations,

☆ Ecologically sensitive sites, Tourist spots,

☆ Major dams, Rivers and Seismic zones,

☆ Past disaster cases and predicted hazards,

☆ Sites of probable mass emergencies (pilgrims *etc.*).

☆ Existing 'Disaster Response Centres.

While mitigation is often taken to mean increasing structural strength, it is more useful to view it within a broader context: mitigation is a management strategy that balances current actions and expenditures with potential losses from future hazard occurrences. More generally, mitigation objectives are those that eliminate or reduce the probability of occurrence of a hazard event, or those that reduce the impacts of hazard occurrence. Preparedness activities are necessary to the extent that other mitigation measures cannot prevent disasters. In the preparedness phase, Governments, organisations and individuals develop, test and maintain plans to save lives and minimise disaster danger (Petak and Atkisson, 1982).

Ecological risk assessment provides a methodology for evaluating the threats to ecosystem function associated with environmental perturbations or stressors and likely to cause a hazard or an increase in the probability (Lowrance and Vellidis, 1995). The objective of the regulatory development activities during the last two decades was to improve our nation's environmental quality. Environmental quality may be defined as a level of measurable ecosystem-health (Gupta, 1996) that is desired, which an organisation's activities and strategies are designed to achieve such with reduction in the risk of a natural hazard or preservation of a geographic area of significance. The scientific and technological information requirements may be classified as: (1) research and development activities aimed to improve the risk assessment process, and (2) innovative disaster prevention approaches and impact minimisation technologies (Rao, 1995). Following are the important aspects to be covered in environmental risk analysis:

1. Identification of potential hazards.
2. Probability of occurrence and prediction/warning,
3. Maximum credible damage scenario,
4. Damage distance analysis,
5. Secondary hazards, and
6. Environmental risk mapping.

Research and Trainig: Approach to Disaster Management Education

Many recent disasters had their main impact in human settlement sites where there is a large concentration of people with severe (direct or indirect) stress on natural factors and dependency of civil infrastructure and services. Environmental degradation has often increased the vulnerability, in particular of low-income groups. Technological risks increasingly play additive role in urban and industrial societies. Increasing carbon depletion (in soil-nature system) is adding to the problem as the increase in risk (frequency X intensity) of hazards like floods, droughts, landslides, and epidemics *etc.*

It is the foremost need that the science of Disaster Management should be brainstormed and conceptualised. A "National Disaster Prevention and prevention and Reduction Agency" should be established under the Ministry of Environment/ Planning and now there is no wisdom in keeping natural and man-made disasters separately while framing the National Policy for Disaster Reduction. The integrated approach should necessarily incorporate the epidemiology, environmental risk analysis/auditing and community psychology/medicine. The disaster management goal should imperatively focus on research priorities of hazard prevention through environmental management and planning, vulnerability reduction and impact minimization.

Organisations involved with emergency response or recovery work view a disaster as providing an opportunity to promote economic development (Bekre, 1994). In particular, collaboration between external donors and local organisations can lead to a strengthening of local organisational capacity to undertake development projects. Recovery communities need to acknowledge the constraints to development posed by natural hazards. In many instances, housing had not failed because the raw material was inadequate, but because they were improperly used. In other cases, population growth, increasing urbanisation, shortage of low-cost land, and impoverishment meant that people incurred severe losses because they were forced to live on marginal land, such as hazardous flood plains in Jamaica (Berke, 1994). Flooding in low lying areas was also intensified because of poorly maintained storm drainage system. Ecologically damaging development practices also compounded the risk, as in the case of deforestation in several countries that induce watershed erosion and flooding, and in tern drought in summer. Relationship between sustainable development and disaster reduction should be recognised as natural disasters can be reduced only if the risks are analysed and taken into account in an overall regional planning. Communities need to have a realistic understanding of the risk to which they are exposed, and have the knowledge, ability, and guidance to take adequate preventive action and protective measures.

Training is a means to improve performance and problem solving capabilities in difficult situations. It is also a continuous process to build capability, in sufficient number of concerned individuals, to bring about planned changes in the system or to plan preventive measures to reduce the risk of a hazard. The disaster management training should involve both preventive and protective approach. The information flow should be based on research on causes of hazards and their reduction. Therefore, in order to fulfil the objectives of IDNDR, training should focus on following issues;

☆ Causes and probability of hazards,

☆ Risk analysis and hazard prevention measures,

☆ Disaster reduction, impact assessment and minimisation,

☆ Disaster preparedness and emergency management,

☆ Loss analysis and mitigation analysis,

☆ Rescue, relief and rehabilitation,

☆ Risk mapping based guidelines.

Disaster Preparedness and Role of Data Management

Many people now accept that human activity itself has created the conditions for disaster events. This partly because of growing awareness that through negligence or inappropriate response, the working of social systems have made a disaster out a situation which otherwise might not have been so serious. There has also been a growth in understanding that it is hazard that are natural, but that for a hazard to become a disaster it has to affect vulnerable people (Cannon, 1993).

Aligning national and local efforts into the adoption of a preparedness and protection strategy may be no easy task. Such a strategy has to make its way into the face of the adopted "relief management policy", and in the face of competing claims of national policies for poverty alleviation, integrated rural development, food security agriculture, economic structure adjustment and environmental protection (Oakley, 1993). Principle problem is that the *"relief-after-the-the-event"* strategy is still being followed as the issue of natural disaster prevention and management is being dealt chiefly by the civil officers and the resource-scientist community is not appropriately made involved in planning better hazard reduction approach. In fact, it requires to establish a National Disaster Management Service. Ministry of Environment and Forests, Government of India has issued a notification (August, 1996) for the constitution of Central Crisis Group, State Crisis Group, District Crisis Group etc. However, it still lacks proper execution. In order to formulate an effective network these agencies may be given following responsibilities:

☆ Help define with communities the nature of their vulnerability

☆ Help ensure the most effective use of scarce resources

☆ To motivate the people on issues of hazard reduction strategy

☆ Encourage the establishment of Local Emergency Preparedness Office

☆ Advice on the undertaking of mitigating actions

☆ Offer training opportunities and manuals

☆ Encourage the application of Environmental Impact Assessment of developmental projects,

☆ Introduction of Strategic Environmental Assessment (of policies) and consequence modeling.

To serve such plans, systems and procedures for disaster prevention and emergency management a comprehensive disaster information system is required. The identification and selection of data for disaster planning has to be based on the availability of reliable information on the range of topics. Existing risk assessment methods assume highly conservative exposure assumptions and toxicity/consequence estimates. By adopting conservative point estimates, the opportunity to quantitatively and more accurately address the input factors as a range estimates (Finkel, 1990). Risk estimates based on *worst-case exposure scenario* do not provide risk managers with lower risks of long-term ecological hazards ultimately leading to environmental/natural disasters (Rao, 1995). The data on these aspects are likely to be well defined, and logically organised whilst other information may not exist, or where available may be both fragmentary and of debious quality. These are the typical difficulties and discrepancies in a disaster planning system for an industrialising society, so within a developing country even wider gaps can easily be imagined in view of their limited professional resources to process and apply data.

Conclusion

Countering natural disasters continues to absorb the efforts of a number of branches of science and engineering and a wide range of institutions, both national and international. It has been estimated that, as a result of natural disasters, during 1970s and 1980s three million lives were lost world-wide, the number of disaster increased threefold, the economic losses per decade almost doubled and the insurance losses per decade quadrupled. In order to achieve the goal of United Nations proclamation of *International Decade for Natural Disaster Reduction (IDNDR)*, nations are required to formulate their policies on *Disaster Reduction* through integrated approach. Emphasis of "Reduction" should be considered primarily on "Prevention" and then on "Impact Minimisation" through preparedness of rescue, relief and rehabilitation. The issues of disaster reduction should be observed in context with the objectives of efficient resource management, water conservation, poverty eradication and environmental protection through public motivation and statutory support.

References

Alexander, D. (1993). Natural Disasters. UCL press, London.

Aysan, Y. F. (1993). Vulnerability assessment. In : Natural Disasters (Ed: P.A. Merriman and C.W.A. Browitt). Thomas Telford, London pp. 1-14.

Berke, P. R. (1994). Sustainable development : Putting the principle in Practice in the Caribbean. DHA News, 7 : 108.

Cannon, T. (1993). A hazard need not a disaster make: Vulnerability and causes of "natural" disasters. In : Natural Disasters (Ed: P.A. Merriman and C.W.A. Browitt). Thomas Telford, London pp. 92-105.

Cornell, J. (1982). The (Great International) Disaster Book. Charles Scribner's Sons, New York.

Finkel, A. M. (1990). Confronting uncertainty in risk assessment - a guide for decision makers. Centre for Risk Management, Resources for the Future, Washington DC.

Gupta, H. K. And J. Combs (1976). Continued seismic activity at the Koyna Reservoir site in India. *Engineering Geology*, **10:** 307-313.

Gupta, A. K. (1996). Forest Fire and Ecosystem-Health : An Ecologists' Overview. *Envoice (A NEERI Journal)*, **6:** 46-49.

Idso, S. B. *et al.* (1875). The dependence of bere soil albedo on soil water content. *Journal of Applied Meteorology*, **13:** 343-347.

Jovanovic, P. (1988). Modeling of Relationship between Natural and Man-made hazards. In : Natural and Man-Made Hazards (Ed: M.I. El-sabh and T. S. Murty). D. Reidel Publishing Company, Boston. pp. 9-20.

Judd, W. R. (Ed) (1974). Seismic effects of reservoir impounding. *Engineering Geology*, **8:** 1-212.

Lowrance, R. And G. Vellidis (1959). A concept Model for Assessing Ecological Risk of Water Quality Function of Bottom Land Hardwood Forests. *Environmental Management*, 19 (2): 239-258.

Mandal, G.S,. (1993). Tropical Cyclones and their Warning Systems. In: Natural Disaster Reduction (Ed: G.K. Mishra and G.C. Mathur), Reliance Publishing House, New Delhi, pp. 128-155.

Oakley, W.J. (1993). A National Disaster Preparedness Service. In: Natural Disasters (Ed: P.A. Merriman and C.W.A. Browitt). Thomas Telford, London. pp. 270-281.

Petak, W.J. and A.A. Atkisson (1982*).* Natural Hazard Risk Assessment and Public Policy. Springer-Verlag, New York.

Rao, V.R. (1995). Risk Management: Time or Innovative Approaches. *Environmental Management*, 19(3): 313-320.

Thampapillai, D.J. and W.F. Musgrave (1985). Flood Damage and Mitigation: a review of Structural and Non-structural Measures and Alternative Decision Frameworks. *Water Resource Research*, 21: 411-424.

2

Climatic Effects of Increasing Concentrations of Atmospheric Methane

R.S. Khoiyangbam and Anil K. Gupta

There is a great concern about global warming due to increasing concentrations of **greenhouse gases** in the atmosphere. Greenhouse gases are the trace gases in the atmosphere which are relatively transparent to the higher energy sunlight, but trap or reflect the lower energy infrared radiation, behaving somewhat like glass in a **greenhouse**. The warming of the earth's atmosphere attribute to the atmospheric trace gases is termed the **greenhouse effect**. The earth's surface would be a frozen mass if it were not for the natural greenhouse effect of the atmosphere. The greenhouse effect is the sum of the interactions between the heat that is attempting to escape from earth to space and the molecules of various radiatively active gases that trap this heat, reradiating it within the atmosphere, and impeding its loss to space. Greenhouse effect is a natural phenomenon, and an essential system for maintaining the earth's temperature. This process occurs largely in the troposphere, in the first 10–15 km of the atmosphere. Without the greenhouse effect, the earth would be 33° C lower than it is, with an average temperature of –18°C.

The history of the earth's climate is characterized by frequent, and sometimes rather sudden, temperature changes. In the past two million years there have been approximately 20 glacial and interglacial periods. In recent earth's history, a relatively steady state balance has been achieved that maintains the average global surface temperature of about 15° C. While the temperature variations in the distant past have been the result of non-anthropogenic forces, the recent change in global climate is largely attributed to human activities. As is now well known, anthropogenic sources of a number of gases such as carbon dioxide, methane, chlorofluorocarbons, nitrogen oxide and ozone are enhancing the greenhouse effect, leading us into a future of uncertain catastrophic global warming. The effects of this would be profound, as polar ice would melt and the oceans would expand with heat, raising sea levels and resulting in large-scale changes to the global climate. In addition, global warming may cause organisms to migrate in order to seek optimal temperatures, or to the changed conditions or face extinction.

Methane is the most abundant organic gas in the atmosphere and second most important anthropogenic greenhouse gas after carbon dioxide. The contribution of carbon dioxide and methane to the present global warming is estimated to be 50 per cent and 20 per cent, respectively. The warming potential of methane is 30 times greater than carbon dioxide in gram per gram basis. Methane is not only a potential greenhouse gas, it also plays an important role in the photochemical reactions of the troposphere and the stratosphere and change of its concentration strongly influences atmospheric chemical reactions. In the troposphere, methane is involved in photochemical reactions that determine the concentrations of ozone and hydroxyl (OH), the 'detergent' of the atmosphere, responsible for the removal of almost all gases that are produced by natural processes and human activities. In the stratosphere, methane's oxidation is an important source of stratospheric water vapour, and thus directly of OH radicals, whose reactions lead to the conversion of ozone destroying NO and NO_2 catalyst to far less reactive nitric acid. On the other hand reactions of OH radical with hydrochloric acid promote the formation of ozone destroying Cl and ClO radicals (Crutzen, 1991).

Table 2.1: Major greenhouse gases and their characteristics.

Gas	Atmospheric Concentration (ppmV)	Annual Concentration Increase (per cent)	Relatively Greenhouse Efficiency ($CO_2 = 1$)	Principal Sources
CO_2	351	0.4	1	Fossil fuel, deforestation
CFC_s	0.00225	5	15, 000	Foams, aerosols
CH_4	1.75	1	25	Wetlands, rice, livestock
N_2O	0.31	0.2	230	Fuels, fertilizers

Source: Flavin (1989).

Current atmospheric methane concentration at 1.75 ppmV is now more than double the pre-industrial (1750–1800) value. At present the atmospheric methane is increasing at the rate of 1.3 per cent per year. These increases in methane are well-correlated with human population growth rates. Human activities such as rice cultivation, livestock rearing, landfills, biomass burning, coal mining, and venting of natural gas or natural-gas pipeline leaks has contributed significantly to the built of atmospheric methane. The production of methane is a biogenic process and is exclusively accomplished by the so called methanogenic bacteria that can metabolize and live only in the strict absence of oxygen. Thus, methanogenic ecosystems are sites, where oxygen-deficient zones develop due to O_2 consumption by respiration and limitation of O_2 diffusion from the atmosphere. The guts of termites or the rumen

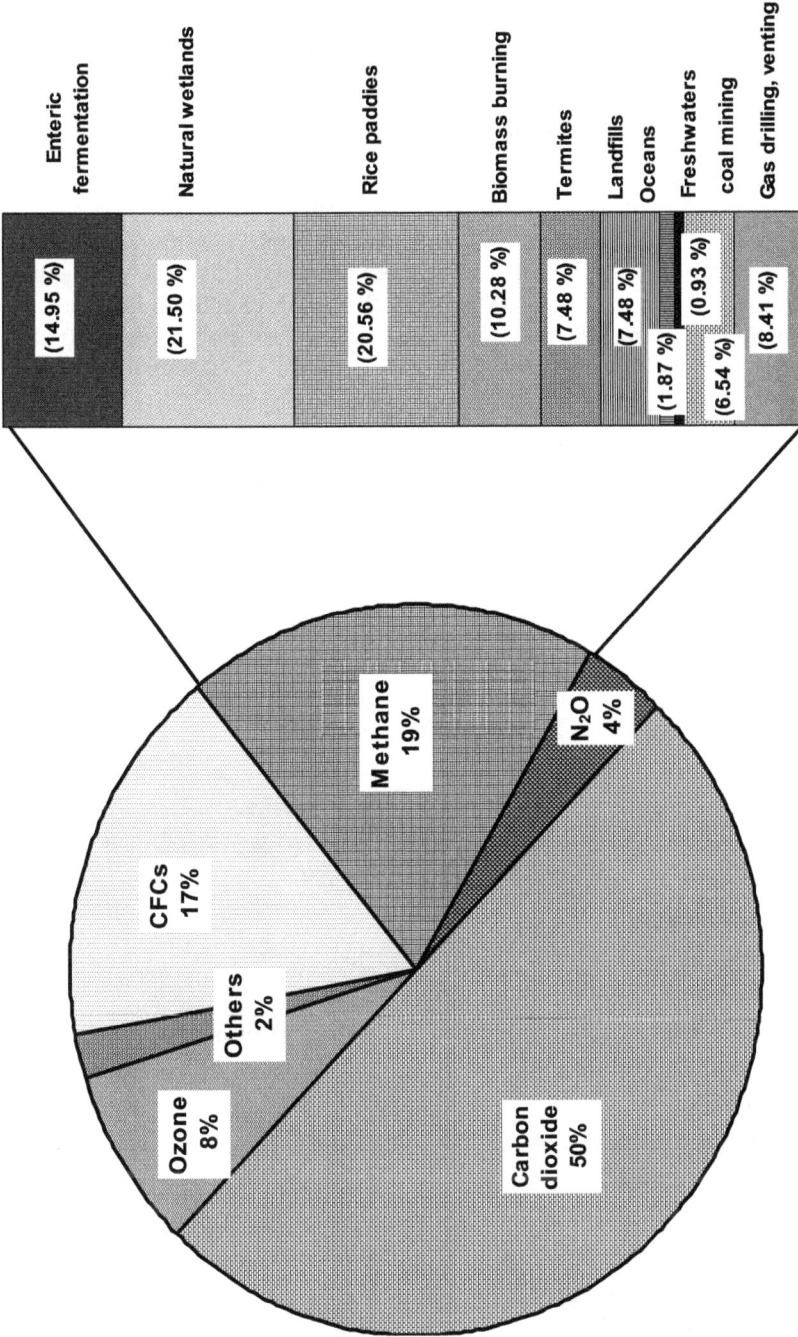

Figure 2.1: Contribution of GHGs to global warming and sources of methane by percentage.

of cattle are also anoxic methanogenic environments, as are the swamps and biogas plants. Once emitted to the atmospheric the concentration of methane is controlled by its reaction with OH in the troposphere, which is the major sink of methane. The next most important sink involves its transport to the stratosphere, where methane is also eventually oxidized to carbon dioxide and water. A third sink for methane is the uptake of methane by microorganisms through oxidation.

Methane: An important gas

Methane is a colourless, odourless, non poisonous atmospheric gas. Its boiling point is −164° C at 760 mm and melting point is −184° C. When liquefied, methane is less dense than water with a relative density of 0.4. It is only slightly soluble in water; 100 ml of water dissolving about 5 ml of methane at 20° C, but is quite soluble in organic liquids such as gasoline, ether, and alcohol. It cannot be liquefies by any pressure at ambient temperature. The critical pressure of methane is 4710 kPA at −

Table 2.2: Some important properties of Methane.

Properties	Value along with Unit
General	
Chemical formula	CH_4
Characteristics	Colourless, odourless
Molecular weight	16.04
Density	0.5576 g/cc
Heat value	1012 BTU/cft at 14.4 psi, 60° F
Critical pressure	684 psi
Critical temperature	− 116° C
Heat capacity	8.0472 cal. deg^{-1}. mole^{-1} at 300° K
Heat of combustion	210.8 kg cal/g. mole at 1 atm, 68° F
Heat of formation	20.34 kg cal/g. mole at 1 atm, 68° F
Related to combustion	
Octane rating	130
Ignition temperature	1202° F
Limits of flammability	5–15 per cent by weight
O_2/CH_4 for complete combustion	3.98 by weight, 2.0 by volume
Products of combustion	$CH_4 + 2O_2 \rightarrow CO_2 + 2 H_2O$
CO_2/CH_4 for complete combustion	2.74 by weight; 1.0 by volume
Related to Greenhouse Effect	
Current atmospheric concentration	1.75 ppmV
Annual concentration increase	1 per cent
Relative greenhouse efficiency ($CO_2 = 1$)	30
Present contribution to global warming	20 per cent

Sources: Lapp (1975) and Khoiyangbam (2002).

82.3°C. Methane burns with a non-luminous flame in air or oxygen, forming carbon dioxide and water. As such, it is the major constituent (up to 97 per cent) of natural gas and forms about 40 per cent by volume of coal gas. It explodes violently when mixed with air (or oxygen) and ignited, and then is believed to the cause of explosion of coal mines, where methane is known as *fire–damp*.

Methane is indeed, one of the simplest of all organic compounds and is by far the most abundant hydrocarbon in the earth's atmosphere. According to one theory, the origins of life go back to a primitive earth surrounded by an atmosphere of methane, water, ammonia, and hydrogen. Energy from the sun, lightning discharges, broke their simple molecule into reactive fragments, these combined to form larger molecules which eventually yield the enormous complicated organic compounds that make up living organisms. The combustible nature of methane has led to its exploitation as a source of heat and light since long (Khoiyangbam, 2003). Biomethanation in anaerobic digesters to produce methane gas as a source of energy is a common phenomena in many countries of the world. Methane is also used in various industries and other applications. By heating methane to 1000°C or by incomplete combustion, it produces carbon in a very finely divided state. Carbon prepared this way is known as *carbon black* and is used to make paints and printer's ink.

Atmospheric Methane Build-Up

The methane concentration in the atmosphere reached about 1.72 ppmV in 1990. It has been increasing at a rate of about 1 per cent on average per year during the century. Pre-industrial methane concentrations have been derived from ice cores. Measurements of the polar ice-core samples from Antarctica and Greenland indicated that, three centuries ago, the concentration of methane in the atmosphere was 650 ppbV (Chappellaz, J., *et al.*, 1990; Stauffer, B., *et al.*, 1988; Rasmussen, R. A., and Khalil, M. A. K., 1984). This is less than half of the 1990's level. According to recent polar ice-core studies the atmospheric methane concentration was only about 350 ppbV during the last glaciation, *i.e.*, ~130 x 10³ years before the present time. The concentration of methane remained virtually unchanged over thousands of years until the beginning of the 18th century. Subsequently, the methane concentration increase slowly at an average rate of 1.5 ± 0.9 ppbV per year between 17000 and 1800, and at a rate of 1.5 ± 1.2 ppbV per year between 1800 and 1900. The trends escalated to 2.3 ± 1.7 ppbV annually between 1900 and 1925, and reached 6.4 ± 3 ppbV per year 1927 and 1956. Between 1962 and 1974, the rate of increase was about 11 ppbV annually. During the decade of 1974–1984, the methane concentration increased at a rate of 17 ± 2.3 ppbV per year (Khalil and Rasmussen, 1987). Currently, between 350 and 600 million tones of methane are released into the atmosphere annually of which approximately 60 million tones of methane accumulate annually in the troposphere after the removal through reactions involving the hydroxyl radical OH (Ciborowski, 1989).

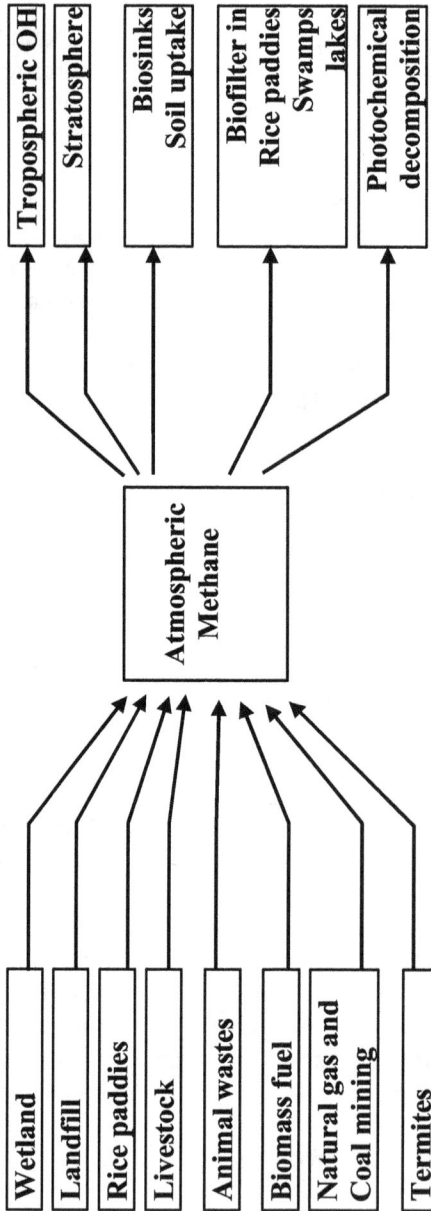

Figure 2.2: Major sources and sinks of atmospheric methane.

Sources of Methane

A source is defined as any process which produces methane and releases it into the atmosphere. Given the concentration of molecular oxygen, carbon dioxide and water, the simultaneous presence of 1.7 ppmV methane in the lower atmosphere with them is an astonishing chemical fact: it constitutes departure from thermodynamic equilibrium greater than 50 orders of magnitude (Watson *et al.*, 1978). It is, therefore, extremely unlikely that a compound containing a carbon atom in its most reduced form would be synthesized within the oxidizing atmosphere. Consequently the search for sources of atmospheric methane has centered on the earth's surface in areas where reducing conditions can be found. There are almost an infinite number of methane sources, but, only a handful release methane in large enough quantities to be significant on a world-wide basis. The basic sources of methane emission to the atmosphere are wetlands, rice paddies, swamps, marshes, shelf, sea, and ocean sediments, biomass burning, ruminants animal, and gas and coal deposit. Anthropogenic processes are responsible for between 55 per cent to 70 per cent of the estimated 600 Tg of methane released annually into the atmosphere (Thorpe, 2008).

Wetlands

'Wetlands' is the collective name for all the areas outside the oceans which are permanently or periodically covered in water. The Ramsar Convention defines wetlands as 'areas of marsh, fen, peatland or water, whether natural or artificial, permanent or temporary, with water that is static or flowing, fresh, brackish, or salt'. The definition includes marine water less than six meters deep, and all rivers and coastal areas, as well as most coral reefs. The world's wetlands represent a large source of methane. Due to the anaerobic fermentation by bacteria in the water-covered soil, they emit 115 million tones of methane annually. Aselmann and Crutzen (1989 and 1990) compiled data on global freshwater wetlands. They grouped the natural wetland ecosystems into six categories: bogs, fens, swamps, floodplains and shallow lakes. Their estimation for the total global natural wetland area is ~5.7 x 10^{12} m². Matthews and Fung (1987) suggested that the tropical/subtropical peat-poor swamps ranging from 20° N to 30° S represent ~30 per cent of the global wetland area and produce ~25 per cent of the total methane emission. Emissions from subtropical and temperate wetlands (20–45° N and 30–50° S) are only 5 Tg yr⁻¹ (4.5 per cent of the total). Northern wetlands (north of 45° N) are calculated to release a total of 38 Tg yr⁻¹ (35 per cent of the total), with 34 Tg yr⁻¹ from wet soils and 4 Tg yr⁻¹ from relatively dry tundra (Milich, L., 1999). Studies in India showed that the annual mean net methane flux from Pulicat lake sediments was 3.7 x 10^9 g yr⁻¹ based on the static measurements (Shalini, *et al.*, 2006). Herbaceous wetland plants plays a significant role in the emissions process, serving both as major carbon source for decomposition and as a transport pathway for the direct release of methane from the anaerobic substrate to the atmosphere bypassing microbial oxidation at the substrate anaerobic interface (Figure 2.3).

Rice Paddies

Rice is the basic food for nearly half the people on earth, most of them concentrated in Asia (IRRI, 1993ₐ). One hundred forty million ha of rice are harvested annually,

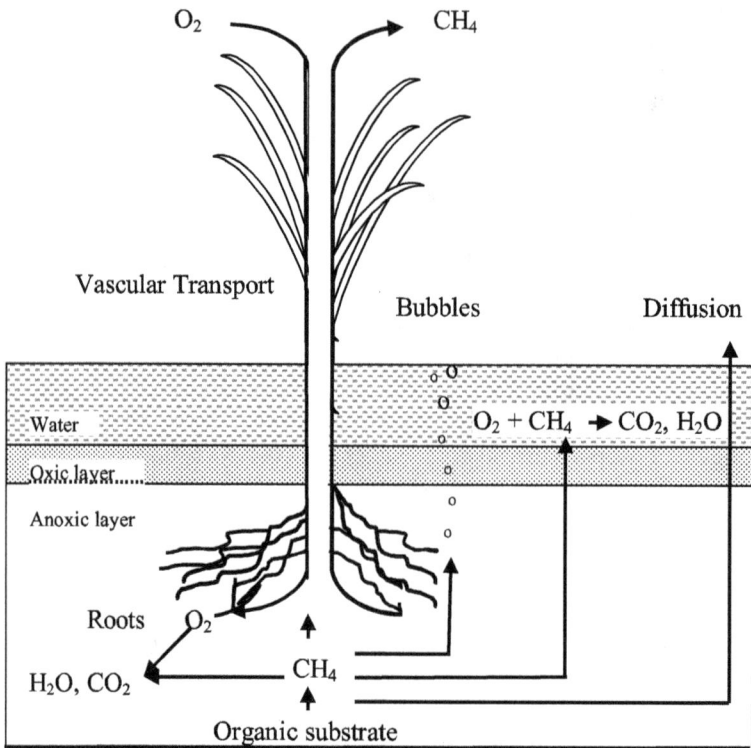

Figure 2.3: CH_4 production, oxidation and transportation in wetland plants.

occupying, about 10 per cent of the arable land world wide. Irrigated rice, which accounts for more than 75 per cent of global rice production, represents one of the main sources of methane. Intensification of rice cultivation to meet the demand for rice by the increasing human population is imperative, especially in Asia where ~90 per cent of the rice is grown and consumed (IRRI, 1993_b). Waterlogged paddy fields are just shallow swamps which are rich in decomposing vegetation and which are distributed in the temperate and tropical zones where temperatures are warmer. Anaerobic conditions exist in the bottom sediments of paddy soils during periods of flooding. The methane production from rice paddy is a function of temperature, sediments and flooding conditions. Recent estimated of the methane source strength of rice fields ranged from 20 to 100 Tg CH_4 per year.

Landfills

Landfills are a significant source of methane, ranking third in anthropogenic sources after rice paddies and ruminants (Peer, *et al.*, 1993). The amount of methane that vent from the landfills is dependent on the physical, chemical and biological components of the soil cover (Tecle, *et al.*, 2008). Landfills, unlike any other anaerobic systems, are extremely heterogeneous environments, characterized by a high solid content. Wastes dumped in the landfill may be categorized into biodegradable and non-biodegradable. Landfill gas mainly comprises of methane and carbon dioxide, with minor amounts of other gases. The gas production potential of a landfill is determined by the quantity and nature of the biodegradable components of wastes. Estimates for worldwide landfill methane emissions range from 9 to 70 Tg yr^{-1} (Bogner and Spokas, 1993). Field and laboratory studies suggest that maximum methane yields from landfilled refuse are about 0.06 to 0.09 m^3 (dry kg)$^{-1}$ refuse, depending on moisture content and other variables, such as organic loading, buffering capacity, and nutrients in landfill microenvironments. At a small number of sites worldwide, landfill methane is flared or recovered commercially for fuel use. At most sites world-wide, the landfill methane is vented into the atmosphere (Figure 2.4).

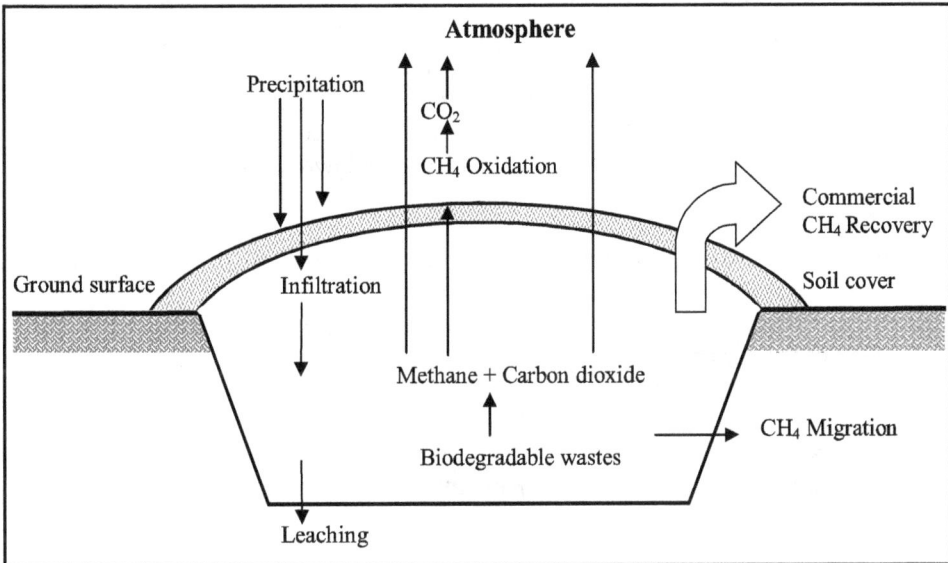

Figure 2.4: CH$_4$ and CO$_2$ emission from landfill.

Termites

Methane has been discovered in the digestive tracts of various insects, including scarab beetles, wood-eating cockroaches and various lower termites. Bacteria in the guts of termites convert most of their wood food into CO$_2$ and CH$_4$. Investigators estimate that there are 3.5 x 10^{17} termites in the world, inhabiting two-thirds of the world's land area and eating one-third of the world's vegetation. Rasmussen and Khalil (1983) estimated the annual global methane production by termites to be 50 million tonnes.

Animals

Domestic livestock are an important component of agriculture in the developing world, and in some cases the only means of production. Methane is a by-product of the microbial breakdown of carbohydrates in the digestive tracts of these herbivores. The largest methane emissions are reported from ruminants, which host large populations of bacteria and protozoan in their rumens. Wild ruminants live entirely on roughage and herbs and the total methane emission by the wild ruminant population of 125 ± 25 million in the subtropical and tropical region has been estimated to be 3 ± 2 million tonnes per year (Crutzen, *et al.*, 1988). Together with the contribution from ruminants in the northern temperate regions, the annual methane production from wild ruminants in the world may, therefore, be equal to 4 ± 2 million tonnes per year. Domestic ruminants, such as cattle and sheep emit large amount of methane as a result of enteric fermentation. Of the domestic animals that emit significant amounts of methane, including a few non-ruminants such as horses and pigs, cattle are by gar the largest methane produces. Currently ruminants contribute ~97 per cent of the annual methane emissions from domestic animals and non-ruminants contributing 3 per cent . In the late 1980s, there were some 1269 million head of cattle worldwide, collectively producing about three-quarters of the total estimated ruminant contribution of 65–100 million tonnes of methane per year. In India, although the emission rate per animal is much lower than the developed countries, due to vast livestock population the total annual methane emission are about 9-10 Tg from enteric fermentation and animal wastes (Sirohi and Michaelowa, 2007).

Biomass Burning

Biomass burning has recently come into focus as a possible source of numerous gases important to atmospheric chemistry. This practice is widespread, especially in the tropics. Fires are set in tropical forests to open lands for agriculture and in tropical savannas to maintain forage quality as well as to prevent colonization by wood plants. In temperate zones, biomass burning is largely the result of deliberately or accidentally set fires and lightning-ignited wildfires. Furthermore, agricultural wastes and fuel wood is combusted as a source of domestic energy, particularly in developing countries. It has been estimated that the annual methane emission from biomass burning is 15-71 million tonnes, which may be about 10 per cent of the global total source.

Natural Gas Leaks, Coal Mining, and Coal Burning

Methane emission into the atmosphere occurs as a direct result of fossil fuel mining and combustion. During the production of oil and natural gas, some portion of the evolved gas is either vented to the atmosphere or flared. On average, the amount of natural gas flared and vented amounts to about 2–3 per cent of global natural gas production. Leaks of gas also occur during the refining and distribution of the gas. It is assumed that about 20–50 million tonnes of methane is released per year globally. During the formation of coal, the methane produced by the decomposition of organic material, becomes trapped in the coal seam. When the coal is mined and processed or crushed, part of the methane is released and adds 6–22 million tonnes to the atmospheric annual budget. Methane is also emitted into the atmosphere as a result of the incomplete combustion of fossil fuels. The rate of emission varies with the type of fossil fuel consumed and the combustion technology used. The fossil fuel combustion adds about 1–2 million tonnes of methane to the atmosphere per year.

Conventional Anaerobic Digesters

In many countries worldwide anaerobic digesters are used to generate methane in the form of biogas as a source of domestic energy. The conventional biogas digesters have exposed areas, from which methane is emitted continuously to the atmosphere (Figure 2..5). The annual contribution per plant to the global methane budget from a fixed dome biogas plant (Capacity 2 m³) operating in plain and hilly region respectively, of northern India amounts to 53.2 kg and 22.3 kg (Khoiyangbam *et al.*, 2004$_a$). There are 10 million biogas pits used in China (Mingxing *et al.*, 1992). In India, more than 3 million family size biogas and 4000 large capacity institutional/community biogas plants have been installed under the National Programme on Biogas Development. This covered only one-fourth of the estimated potential of 12 million biogas plants installation in India. In future the number of biogas plants is going to rise considerably, thereby increasing the contribution of methane to the atmosphere (Khoiyangbam, *et al.*, 2004b).

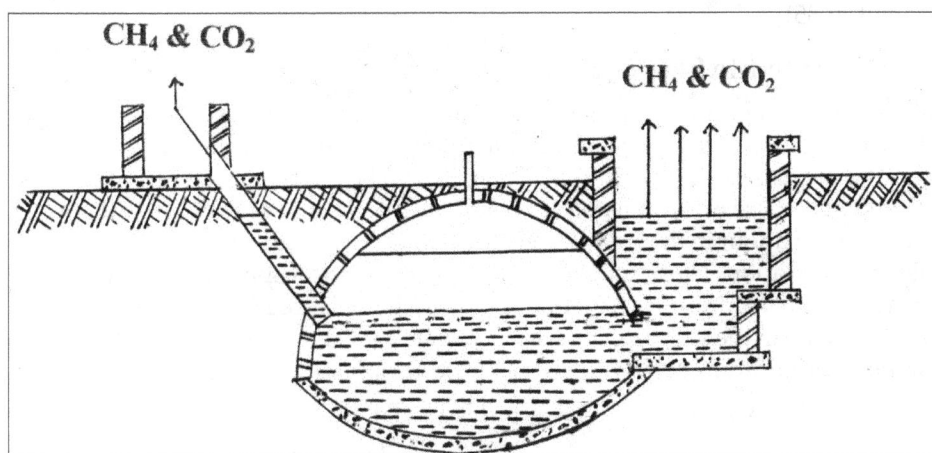

Figure 2.5: CH_4 and CO_2 emission from conventional biogas plant.

Sinks for Atmospheric Methane

The increase of atmospheric methane concentration results from an imbalance between methane production and oxidation. The concentration of methane is the atmosphere is incommensurably small when compared to that found in soils and sediments where methane is generated. Methane is removed from the atmosphere primarily by oxidation within the troposphere. This process occurs in several stages and is initiated by the reaction with the hydroxyl radicals. The next most important sink involves its transport to the stratosphere, where methane is also eventually oxidized to carbon dioxide and water. A third sink for methane is the uptake of methane by microorganisms through oxidation.

Reaction with the Hydroxyl (OH) Radical

The atmospheric concentration of methane is controlled by its reaction with OH in the troposphere, which is the major sink of methane. Approximately 90 per cent of the tropospheric budget of methane is destroyed by reaction with the hydroxyl radicals, leading to methane's short atmospheric residence time of about 10 years. About on quarter of the tropospheric budget of CO us a product of methane's photochemical oxidation. The hydroxyl radical required to initiate the oxidation of methane is produced via the photolysis of tropospheric ozone by solar ultraviolet–radiation of short wavelength less than 310 nm. The resulting excited oxygen atom then reacts with water vapour to form the hydroxyl radical,

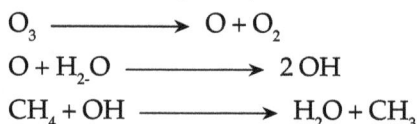

$$O_3 \longrightarrow O + O_2$$
$$O + H_2O \longrightarrow 2\,OH$$
$$CH_4 + OH \longrightarrow H_2O + CH_3$$

It has been estimated that the hydroxyl radical concentration in troposphere is 6.5×10^5 molecules cm^{-3} (Volz, *et al.*, 1981). The resulting destruction rate of methane in the troposphere was, therefore, calculated to be 500 ± 230 million tones per year (Ehhalt, 1985).

Transport to the Stratosphere

About 10 per cent of the tropospheric methane escapes destruction in the lower atmosphere and enters into the stratosphere. Once it is transported into the stratosphere, it is oxidized forming CO_2 and H_2O. The global transfer rate of methane to the stratosphere was estimated as about 50 ± 20 million tones per year. About half of the water vapour in the stratosphere is estimated to have originated as a result of methane oxidation. These could increases the polar stratospheric clouds that have been linked to the dramatic ozone depletion in the polar regions. The reaction of Cl with methane in the stratosphere has, however, been recognized as one of the sinks for the chlorine atoms, which are the major stratospheric one depleter.

$$CH_4 + Cl \longrightarrow HCl + CH_4$$

Hydrochloric acid, which is a stable compound in the lower stratosphere, diffuses into the troposphere, where it is removed by precipitation.

Methane Oxidation

Methane oxidation by microorganisms is one of the important factors, which influences methane flux into the atmosphere. Methane emission from a particular ecosystem is basically controlled by two different microbial processes by methane production and methane oxidation. Only the part of methane, which is not oxidized, will enter the atmosphere. Methane oxidation takes place in many ecosystems, aquatic and terrestrial wherever stable sources of methane are available. Oxidations of methane are known in soils of desert, savanna, temperate and sub–arctic forest, peat, tundra, agricultural, hyper saline, etc. Global soil methane oxidation rates have been estimated to range from 5 to 50 Tg per year, representing 1 to 10 per cent of current estimated of the methane flux of the atmosphere (Crutzen, 1991). Jang *et al.* (2006) based on the data mined from a review of literature determined t hat mean methane oxidation rate for forest soils was 1.90 mg m^{-2} d^{-1}.

Microbiology of Methane Production

The formation of methane is a unique biological event which is confined to a small group of bacteria. These organisms are poorly understood and are difficult to obtain in pure culture. Yet the biological formation of methane is common in nature, and methanogenic bacteria are readily found in anaerobic environments where organic matter is being vigorously decomposed. In these environments they are terminal organisms in the microbial food chain.

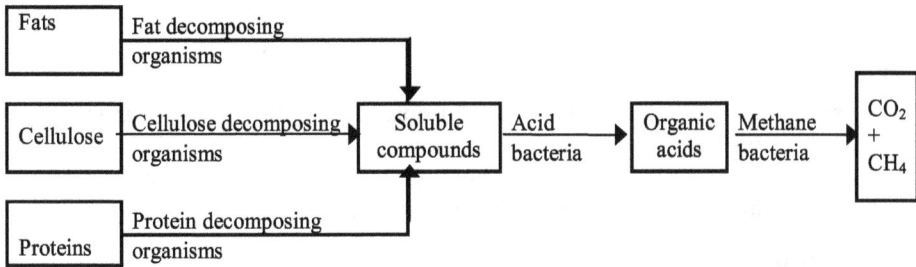

Figure 2.6: Anaerobic bioconversion of organic matter to methane.

Despite the difference in the organic matter and 'specialist' microorganisms involved, anaerobic methanogenic degradation requires the co-operation of four types of bacteria within a substrate food chain. They are the hydrolytic and fermenting bacteria, H^+–reducing bacteria, homoacetogenic bacteria and methanogenic bacteria. These microorganisms participating in the anaerobic breaking down are strict anaerobes. The non-methane producing group consists of several bacteria, fungi and protozoa. The bacteria are the largest in number and consist of fermentative, acetogenic and homoacetogenic group.

Table 2.3: Methane bacteria and their utilizable substrates and products

Bacterium	Substrate	Products
Methanobacterium formicum	CO, H_2 + CO_2 and formate	CH_4
M. mobilis	H_2 + CO_2 and formate	CH_4
M. propionicum	Propionate	CH_4, CO_2 + acetate
M. ruminantium	H_2 + CO_2 and formate	CH_4 + CO_2
M. soehngenii	Acetate and butyrate	Propionate + acetate
M. suboxydans	Caproate and butyrate	CH_4 + CO_2
Methanococcus mazei	Acetate and butyrate	CH_4
M. vannielli	H_2 + CO_2 and formate	CH_4
Methanosarcina barkerrii	H_2 + CO_2 and methanol	CH_4 + CO_2
M. methanica	Acetate and butyrate	CH_4 + CO_2

Source: Chawla (1985)

Factors Influencing Methanogenesis

Methanogenic Substrate

Methane is produced in nature from a limited range of substrate: acetate, formate, H_2–CO_2, and possibly methanol or some related intermediate (Table 2.3). Since the organic matter that enters the methanogenic ecosystem such as lake sediments is mainly protein, carbohydrate, and lipid polymers or heteropolymers, methanogenic bacteria are obligately associated with other bacteria that ferment these compounds to acetate, formate, and H_2–CO_2.

pH

Methanogens are pH sensitive populations. Most of them grow over a relatively narrow pH range (about 6 – 8) and the optimal pH is about 7. Nevertheless, methane production does occur in acidic environments such as pet bogs. There are also a few strains of alkaliphilic bacteria that produce methane.

Temperature

Soil temperature and methane production or emissions generally have a positive relationship.

Soil Redox Potential

Redox potential measures the ability of a soil environment to supply electrons to an oxidizing agent, or to take up electrons from a reducing agent. In a complex environment such as a soil ecosystem, redox potential is determined by a number of reactions, and is a comprehensive measurement of soil chemical and biochemical oxidation–reduction process. The critical Eh for methane production has recently been observed in the range of –140 to –160 mV. An exponential relationship between methane production and soil Eh observed when soil Eh was lower than –150 mV.

Fertilization

Fertilization, especially application of N, is essential for intensive rice cultivation. Fertilizer application alters soil pH, microbial populations and plant litter and root exudates inputs, affecting methane production. The application of fertilizers such as $(NH_4)_2 SO_4$, and $NH_4 NO_3$ is known to produce a gradual suppression of methane production. Nitrate tends to buffer soil Eh and slow methane production. An increase in soil Eh from –200 to +200 mV was observed in a laboratory study after nitrate was applied to a highly reduced soil. This increased soil Eh to much higher than the critical soil Eh for methane production.

Microbiology of Methane Consumption

Methanotrophs are the microorganisms which oxidized methane into carbon dioxide and water. Methanotrophs belong to gram-negative Eubacteria of the family Methanococcaceae. All are a subset of the larger group of organisms the methylotrophs, having the ability to use compounds other than CO_2 containing one or more carbon atoms but no carbon–carbon bonds as the sole carbon source of growth (Dudey *et al.*, 1996). These organisms are strict aerobes which besides methane are shown to oxidized methanol, dimethylether, carbon monoxide, bromoethane, ethylene, propylene, etc. Facultative methanotrophs, which could grow on multi-carbon compounds, were also described. Methanotrophs have been classified into a few main genera which are shown in Table 2.4.

Table 2.4: Certain characteristic features of Methanotrophic bacteria from soils, sediments and water.

Genus	Cell Structure	Substrate Utilized for Growth	Obligate Methanotrophy	Resting Structure
Methylococcus	Non-motile coccus	CH_4, CH_3OH	+	Cyst
Methylomonas	Polarly flagellated rod	CH_4, CH_3OH	+	Cyst
Methylobacter	Polarly flagellated rod	CH_4, CH_3OH	+	Cyst
Methylocystis	Non-motile	CH_4, CH_3OH	+	Cyst
Methlyosinus	Polarly flagellated rod Often pear shaped	CH_4, CH_3OH	+	Exospore
Methylobacterium	Polarly flagellated rod	CH_4, CH_3OH	–	Exospore

Source: Khoiyangbam (2002).

Factors Influencing Methane Oxidation

Methane Availability

Methane concentration affects the rate of consumption of methane by methanotrophs both directly and indirectly. Methane oxidation at high and low concentration is performed by different types of methanotrophs. The overall methane oxidation activity increased at elevated concentrations.

Oxygen Availability

Methane oxidizing bacteria require O_2 for metabolism. All the methanotrophs so far isolated and described need O_2 and unable to use other electron acceptors. However, there is convincing evidence that methane is partially oxidized by anaerobic processes in anoxic marine sediments and hyper saline water.

Water Potential

Water potential is the free energy of water in a system relative to the free energy of a reference pool of pure, free water. Uptake rates of atmospheric and elevated methane decreased when water potential were reduced by either adding ionic or nonionic solutes to a soil of a fixed content.

Temperature

Soil methane consumption is much less sensitive to changes in temperature than other biological processes.

Nitrogenous Compounds

The influence of inorganic N influence on methane oxidation is due to the shifts in the population structure and the kinetics of methanotrophs. Methane consumption in mineral soils is significantly inhibited by addition of either ammonium or nitrate.

pH

In general, both low and elevated pH has an inhibitory effect on methane consumption although the mechanism responsible for this effect is not known.

Algae

Methane emission decreased substantially in the presence of thin layer of algae microcosm. Decrease of methane has been resulted either as a physical barrier to diffusion of methane or to O_2 released or subsequent stimulation of methanotrophs or both the factors attributed to the lower flux.

Effects of Global Warming

Many predictions are being made about the changes in world climate as a result of global warming. Scientists do it by constructing computer models. The reliability of the conclusions deduced from such models is not being without questions. However, scientist agree that the rate of global warming for the past quarter century was greater than any previous period since 1880, and the warming would further raise the temperature of earth by 3–5° C if increase in CO_2 doubles. On a world wide average the earth's temperature would rise by 1 to 3.5° C by the year 2100.

Climatic Changes

Besides change in global temperature, there would be a shift in rainfall pattern. Large areas of the world will experience climatic shocks: some regions may lose up to 30 per cent of their annual rainfall – with concomitant problems relating to agriculture, food supply, forest health, soil moisture, drinking water supply, and public health. Other areas may gain 60 per cent in annual rainfall – with problems caused by storm damage, agricultural loss, flash flooding, and implications for public health. Deserts will receive more rainfall and traditional agricultural land will receive less rainfall. There will be pole ward extension of deserts.

Sea Level Rise

The temperature of the oceans would increase, leading to some melting of polar ice and mountain ice caps, which could result in a rise in sea levels around the world. Surface ocean temperatures have already risen since the beginning of the century by between 0.3 and 0.9° C, an increase that correlates to the warming of the land. When the sea rises, coastal areas are eroded of flooded. A one-meter sea level rise would flood over 10 per cent of Bangladesh. Countries like Egypt and India would suffer on a widespread basis. Half the Commonwealth countries in the Pacific – all small island nations – are at risk. Some would disappear.

Climate Refugees

Forced by starvation, agricultural loss, flooding, or drought, large numbers of people would be moving within countries and between them and the socio-economic structure of the world will be crumbled down. The United Nations is already considering what to do with the world's 'climate refugee'. In addition, global warming may cause organisms to migrate in order to seek optimal temperatures (organisms tracking), or to the changed conditions or face extinction.

Human Diseases

Human health may also be affected as rising temperature expands the areas vulnerable to tropical diseases such as malaria, dengue, filariasis, etc. One of the biggest threats foreseen is the influence of temperature and rainfall on distribution of such disease carrying insects.

Global Warming's Impact on Methane Fluxes

Climate change and the subsequent associated effects can influence microbial processes and cause further changes. Increases of even a few degrees in seasonally averaged temperatures will increase the metabolic activities of methanogenic processes noticeably. The wet tundra ecosystem under increased temperatures and thaw depth will emit more methane. Increase in precipitation for northern latitudes will continue to waterlogged the soil expanding substantially methane emissions sites. There is no lack of organic material for methanogenic bacteria; Artic ecosystem contains approximately 49.7 billion tonnes of carbon as dead organic matter in the active soil layer and upper permafrost, and wet tundra ecosystems are reported to accumulate 30 Tg of carbon annually (Milich, 1999).

Mitigation Strategies

It is just not methane that contributes to the greenhouse effect. There are other pollutants that are credited with contributing to global warming. Since the sources of greenhouse gases are widespread ranging from transportation to electricity generation to gas to oil to agriculture to deforestation and the solution to global warming lies not in mitigation of single source, but a multi-prong approach to reduce all the major sources. A source wise mitigation strategy to minimized methane emission has been discussed below.

Wetlands

While the global trend of loss on natural wetlands has reduced methane emissions, this has been countered by increases in paddy-rice cultivation. There are no suitable mitigation strategies to limit methane emission from natural wetlands, which are already endangered ecosystem worldwide and it is reasonable to focus attempts to mitigate on artificial wetland such as rice cultivation. A promising approach is in the modification of flooding patterns in irrigated rice cultivation effectively inducing short term 'droughts'. Application of green manures, composts and straw enhance methane emissions. Allowing the organic wastes to compost aerobically prior to its incorporation into the soil reduces methane emissions. Changing cultivation techniques also affects methane emissions. Since most methane released from rice paddies is through the plants' micropores growing high yielding rice cultivars with low root exudation not only will likely reduce methane fluxes to the atmosphere but also have the potential for higher yields.

Animals

The use of forages with high solubility carbohydrates reduces the level of hydrogen available for methane production. For human subjects, sulfate availability in the diet decreases methane production provided that sulfate reducing bacteria are present in the gut. In manure pits, dry manure produces only one-seventh the methane compares to wet manure. Provision of drying methods is therefore a useful strategy.

Landfills

Landfill methane should be recovered for fuel use. If recovery deemed inappropriate it should be flared, instead of venting into the atmosphere.

Biomass Burning

Forest fires, slash-and-burn agriculture and diversion of forest land for agriculture in tropics should be restricted. High efficiency wood stoves should be popularized in the developing countries. Fuel substitution is another option.

Fossil Fuel

Gas industries should take care to avoid gas leakage during extraction and distribution. When needed during the normal course of operation the gas should be flared off rather than venting it. Replacement of coal as an industrial and domestic fuel with other forms of energy should be encouraged.

Growing of Algae as Biofertilizer

The presence of algae on the surface of a flooded soil reduces methane emission through methane oxidation. In the biogas digested slurry, the presence of natural algal cover on the surface enhanced the oxidation rate of methane by 21–94 per cent of the produced methane (Khoiyangbam, 2002). Algae are used as valuable biofertilizer in many countries. Encouraging this biofertilizer amongst the farmers in the rice cultivated zones will not only augment the nutrient supply but also arrest the methane budget from rice fields to the atmosphere.

Conclusion

There is a strong consensus amongst atmospheric scientist that global warming during the next few decades will be intensified as a result of the accumulation of greenhouse gases in the atmosphere. The consequences of this temperature increase could be extremely disturbing. Methane is a strong greenhouse gas with a significantly averse environmental impact. It controls numerous chemical processes and species in the troposphere and stratosphere. The methane quantities attributed to individual sources vary in the assessments of many authors. However, the majority of researchers agree that the basic sources of methane emission to the atmosphere are wetlands, termites, natural forest fires and enteric fermentation in wild ruminants as natural sources. Rice cultivation, livestock farming, landfills, biomass burning, biomethanation for energy, sewage disposal, coal mining, and venting of natural gas and pipeline leaks as the anthropogenic sources. The increase of atmospheric methane concentration results from an imbalance between methane production and oxidation in the methane sinks. The troposphere is the largest sink, in which about 90 per cent of atmospheric methane is oxidized. About 10 per cent of the methane emitted at the surface of the earth rises into the stratosphere, where it is also oxidized. Methane oxidation is one of the factors which substantially influence methane flux to the atmosphere. Methanotrophic bacteria are a powerful 'filter' of methane on its path from the sites of its generation to its escape to the atmosphere. Rapid human population explosion and their activities have been correlated with increases in atmospheric methane as well as other major greenhouse gases. About 50 per cent of the present methane sources are controlled by mankind. Since there are less suitable mitigation strategies to limit methane emission from natural sources, it is therefore reasonable to focus attempts to mitigate on methane sources controlled by human activities. Source based, appropriate mitigating strategies need to be developed to minimize the emission rates for all these significant anthropogenic methane sources.

References

Aselmann, I., and Crutzen, P. J. (1989). Global distribution of natural freshwater wetlands and rice paddies, their net primary productivity, seasonality and possible methane emissions. *J. Atmospheric Chemistry*. **8**: 307–58.

Aselmann, I., and Crutzen, P. J. (1990). A global inventory of wetland distribution and seasonality, net primary production and estimated methane emissions. In *Soils and the Greenhouse Effect*, ed. Bouwman, A. F., John Wiley, Chichester. pp. 441–9.

Bogner, J., and Spokas, K. (1993). Landfill methane: Rates, fates, and role in global carbon cycle. *Chemosphere*. **26(1–4)**: 369–86.

Chappellaz, J., Barnola, J. M., Raynaud, D., Korotkevich, Y. S. and Lorius, C. (1990). Ice-core record of atmospheric methane over the past 160 000 Years, *Nature*, **345**: 127–31.

Chawla, O. P. (1986). Advances in Biogas Technology. Pub. and Inf. Div., Indian Council of Agricultural Research, New Delhi.

Ciborowski, P. (1989). Sources, sinks, trends and opportunities. In *The Challenge of Global Warming*, Ed. D. E. Abrahamson. Island Press, Washington, DC. pp. 248–55.

Crutzen, P. J. (1991). Methane's sinks and sources. *Nature*. **350**: 380–81.

Crutzen, P. J., Aselmann, I., and Seiler, W. (1988). Methane production by domestic animals, wild ruminants, other herbivorous fauna and human. *Tellus*. **38B**: 271–84.

Dubey, S. K., Kashyap, A. K., and Singh, J. S. (1996). Methanotrophic bacteria, methanotrophy and methane oxidation in soil and rizhosphere. *Tropical Ecology*. **37(2)**: 167–82.

Ehhalt, D. H. (1985). On the rise, Methane in the global atmosphere. *Environment*, **27(10)**: 6–12.

Flavin, C. (1989). Slowing global warming: A worldwide strategy, *Worldwatch Paper 91*, Worldwatch Institute, Washington, DC.

IRRI–International Rice Research Institute. (1993$_a$). IRRI Rice Almanac, Manila, Philippines, pp.142.

IRRI–International Rice Research Institute. (1993$_b$). Rice research in a Time of change, Manila, Philippines, p.79.

Jang, I., Lee, S., Hong, Ji-Hyung and Kang, H. (2006). Methane oxidation rates in forest soils and their controlling variables: a review and a case study in Korea. *Ecol Res.* 21: 849-854.

Khalil, M. A. K. and Rasmussen, R. A. (1987). Atmospheric methane: Trends over the last 10, 000 years. *Atmospheric Environment,* **21:** 2445–52.

Khoiyangbam, R. S. (2002). Evaluation of greenhouse gases emission from conventional biogas plants and manure pits under varying climatic conditions of India, Ph. D Thesis, IARI, New Delhi.

Khoiyangbam, R. S., Sushil Kumar, and Jain, M. C. (2004$_b$). Methane losses form floating gasholder type biogas plants in relation to global warming. *Journal of Scientific and Industrial Research.* **63:** 344–47.

Khoiyangbam, R. S., Sushil Kumar, Jain, M. C., Arun, K., and Vinod, K. (2003). Methane emission form community biogas plant at Masudpur, Delhi. *Current Science.* **84(4):** 499–501.

Khoiyangbam, R. S., Sushil Kumar, Jain, M. C., Gupta, N., and Arun, K. (2004$_a$). Methane emission from fixed dome biogas plants in hilly and plain regions of northern India. *Bioresource Technology.* **95:** 35–39.

Lapp, H. M. (1975). Methane production from animal wastes: Fundamental Consideration, *Canadian Agricultural Engineering,* **17(2):** 101

Metthews, E. and Fung, I. (1987). Methane emission from natural wetlands: Global distribution, area and environmental characteristics of sources. *Global Biogeochemical Cycles.* **1(1):** 61–86.

Milich, L. (1999). The role of methane in global warming: where might mitigation strategies be focused? *Global Environmental Change.* **9:** 179–201.

Mixing, W., Aiguo, D., Xingjian, S., Lixin, R., Renxing, S., Schutz, H., Seiler, W., Rasmussen, R. A., and Khalil, M. A. K. (1993). Sources of Methane in China. Asian Workshop – cum – Training Course on methane emission studies, September 20 – 24, 1993. National Physical Laboratory, New Delhi.

Moiser, A. R., Schimel, D. S. (1997). Influence of agricultural nitrogen on atmospheric methane and nitrous oxide. *Chemistry and Industry.* **23:** 874–77.

Mudge, F., and Adger, W. N. (1995). Methane fluxes from artificial wetlands a global appraisal. *Environmental Management.* **19(1):** 39–55.

Peer, R. L., Thorneloe, S. A., and Epperson, D. L. (1993). A comparison of methods for estimating global methane emissions from landfills. *Chemosphere.* **26(1–4):** 387–400.

Rasmussen, R. A. and Khalil, M. A. K. (1983). Global production of methane by termites. *Nature,* **301:** 700–2.

Rasmussen, R. A. and Khalil, M. A. K. (1984). Atmospheric methane in the recent and ancient atmosphere: Concentrations, trends and inter hemisphere gradient. *J. Geophys. Res.,* **89:** 11599–605.

Shalini., A., Ramesh, R., Purvaja, R. and Barnes, J. (2006). Spatial and temporal distribution of methane in an extensive shallow estuary, south India. *J. Earth Sys. Sci.* 115 **(4):** 451-460.

Sirohi, S. and Michaelowa, A. (2007). Sufferer and cause: Indian livestock and climate change. *Climatic change.* 85: 285-298.

Stauffer, B., Lochbronner, E., Oeschger, H. and Schwander, J. (1988). Methane concentration in the glacial atmosphere was only half that of the preindustrial Holocene. *Nature,* **332:** 812–14.

Tecle, D., Lee, J. and Hasan, S. (2008). Quantitative analysis of physical and geotechnical factors affecting methane emission in municipal solid waste landfills. *Environ. Geol.* DOI 10.1007/s00254-008-1214-3.

Thorpe, A. (2008). Enteric fermentation and ruminant eructation: the role (and control?) of methane in climate change debate. *Climatic change.* DOI 10.1007/s10584-008-9506-x

Volz, A., Ehhalt, D. H., and Derwent, R. G. (1981). Seasonal and latitudinal variation of ^{14}CO and the tropospheric concentrations of OH radicals. *J. Geophys. Res.,* **86:** 5163–71.

Wang, Z. P., Crozier, C. R., and Patrick, W. H. (1994). Methane emission in flooded rice soil with and without algae. *Advances in Soil Science, soil management and greenhouse effect.* Lewis Publishers, U. S. A. pp. 245–50.

Watson, A., Lovelock, J. E., and Margulis, L. (1978). Methanogenesis, fires and the regulation of atmospheric oxygen. *BioSystem.* **10:** 293–98.

3

Coastal and Sea Erosion: Risk Profile and Mitigation Strategies

Sushma Guleria, Sreeja S. Nair and Anil K. Gupta

Coastal/Sea Erosion?

Every land mass on Earth has miles of coast at the interface between the hydrosphere and the lithosphere. Natural forces such as wind, waves and currents are constantly shaping the coastal regions. The combined energy of these forces moves land materials.

The landward displacement of the shoreline caused by the forces of waves and currents is termed as *coastal erosion*. It is the loss of sub-aerial landmass into a sea or lake due to natural processes such as waves, winds and tides, or even due to human interference. While the effects of waves, currents, tides and wind are primary natural factors that influence the coast the other aspects eroding the coastline include: the sand sources and sinks, changes in relative sea level, geomorphological characteristics of the shore and sand, etc. other anthropological effects that trigger beach erosion are: construction of

Glossary of Terms

☆ **Coastline** – is the interface between the ocean and the land - dynamic morphological entity

☆ **Erosion** – is the wearing away of land by the action of natural forces. On a beach, the carrying away of beach material by wave action, tidal currents, littoral currents, or by deflation

☆ **Wave** - A ridge, deformation, or undulation of the surface of a liquid (sea water)

☆ **Littoral drift** - The movement of sediment along-shore

artificial structures, mining of beach sand, offshore dredging, or building of dams or rivers.

Causes of Erosion?

Coastal erosion occurs when wind, waves and long shore currents move sand from the shore and deposits it somewhere else. The sand can be moved to another beach, to the deeper ocean bottom, into an ocean trench or onto the landside of a dune. The removal of sand from the sand-sharing system results in permanent changes in beach shape and structure. The impact of the event is not seen immediately as in the case of tsunami or storm surge. But it is equally important when we consider loss of property. It generally takes months or years to note the impact of erosion; therefore, this is generally classified as a *"long term coastal hazard"*.

Major Causes of Coastal Erosion

I.   Natural Causes

1.   Action of Waves

Waves are generated by offshore and near-shore winds, which blow over the sea surface and transfer their energy to the water surface. As they move towards the shore, waves break and the turbulent energy released stirs up and moves the sediments deposited on the seabed.

2.   Winds

Winds acts not just as a generator of waves but also as a factor of the landwards move of dunes (Aeolian erosion).

3.   Tides

Tides results in water elevation to the attraction of water masses by the moon and the sun. During high tides, the energy of the breaking waves is released higher on the foreshore or the cliff base (cliff undercutting). Macro-tidal coasts (*i.e.* coasts along which the tidal range exceeds 4 meters), all along the Atlantic sea (*e.g.* Vale do Lobo in Portugal), are more sensitive to tide-induced water elevation than micro-tidal coasts (*i.e.* tidal range below 1 meter).

4.   Near-shore Currents

Sediments scoured from the seabed are transported away from their original location by currents. In turn the transport of (coarse) sediments defines the boundary of coastal sediment cells, *i.e.* relatively self-contained system within which (coarse) sediments stay. Currents are generated by the action of tides (ebb and flood currents), waves breaking at an oblique angle with the shore (long-shore currents), and the backwash of waves on the foreshore (rip currents). All these currents contribute to coastal erosion processes.

5.   Storms

Storms result in raised water levels (known as storm surge) and highly energetic waves induced by extreme winds (Cyclones). Combined with high tides, storms may result in catastrophic damages such as along the east coast of India (Orissa Super Cyclone, 1999). Beside damages to coastal infrastructure, storms cause beaches and dunes to retreat of tenths of meters in a few hours, or may considerably undermine cliff stability.

6. Catastrophic Events

In addition to the daily, slow sculpting of the coast, other events like tsunamis which result in major coastal changes over very short time periods. These are referred to as catastrophic events because of the extensive damage that is caused and the unpredictable nature of the event.

7. Slope Processes

The term "slope processes" encompasses a wide range of land-sea interactions which eventually result in the collapse, slippage, or topple of coastal cliff blocks. These processes involve on the one hand terrestrial processes such as rainfall and water

8. Vertical Land Movements (Compaction)

Vertical land movement – including isostatic rebound, tectonic movement, or sediment settlement – may have either a positive or negative impact on coastline evolution. If most of northern Europe has benefited in the past from a land uplift (*e.g.* Baltic sea, Ireland, Northern UK)

9. Sea Level Rise

Sea level has risen about 40 cm in the past century and is projected to rise another 60 cm in the next century. Sea level has risen nearly 110 meters since the last ice age. Due to global warming, average rise of sea level is of the order of 1.5 to 10 mm per year. It has been observed that sea level rise of 1 mm per year could cause a recession of shoreline in the order of about 0.5 m per year.

II. Anthropogenic Causes

Human influence, particularly urbanisation and economic activities, in the coastal zone has turned coastal erosion from a natural phenomenon into a problem of growing intensity. Anthropological effects that trigger beach erosion are: construction of artificial structures, mining of beach sand, offshore dredging, or building of dams or rivers. Human intervention can alter these natural processes through the following actions:

- ☆ Dredging of tidal entrances
- ☆ Construction of harbours in near shore
- ☆ Construction of groins and jetties
- ☆ River water regulation works
- ☆ Hardening of shorelines with seawalls or revetments
- ☆ Construction of sediment-trapping upland dams
- ☆ Beach nourishment

☆ Destruction of mangroves and other natural buffers

☆ Mining or water extraction

Coastal Erosion in India

The Indian coastline is about 7517 km, about 5423 km along the mainland and 2094 km the Andaman and Nicobar, and Lakshadweep Islands . The coastline comprises of headlands, promontories, rocky shores, sandy spits, barrier beaches, open beaches, embayment, estuaries, inlets, bays, marshy land and offshore islands. According to the naval hydrographic charts, the Indian mainland consists of nearly 43 per cent sandy beaches, 11 per cent rocky coast with cliffs and 46 per cent mud flats and marshy coast. Oscillation of the shoreline along the Indian coast is seasonal. Some of the beaches regain their original profiles by March/April. Fifty per cent of the beaches that do not regain their original shape over an annual cycle undergo net erosion. Shoreline erosion in the northern regions of Chennai, Ennore, Visakhapatnam and Paradip ports has resulted due to construction of breakwaters of the respective port. At present, about 23 per cent of shoreline along the Indian main land is affected by erosion.

Table 3.1: Types of coastal erosion in maritime states of India.

State	Sandy Beach (Per cent)	Rocky Coast (Per cent)	Muddy Flats (Per cent)	Marshy Coast (Per cent)	Total Length* (km)	Length of Coast Affected by Erosion** (km)
Gujarat	28	21	29	22	1214.7	36.4
Maharashtra	17	37	46	6526	263.0	
Goa	44	21	35	—	151.0	10.5
Karnataka	75	11	14	—	280.0	249.6
Kerala	80	5	15	—	569.7	480.0
Tamil Nadu	57	5	38	—	906.9	36.2
Andhra Pradesh	38	3	52	7	973.7	9.2
Odisha	57	—	33	10	476.4	107.6
West Bengal	—	—	51	49	157.5	49.0
Daman and Diu					9.5	—
Pondichenry					30.6	6.4
Total mainland	43	11	36	10	5422.6	1247.9
Lakshadweep					132.0	132.0
Andaman and Nicobar					1962.0	—
Total					7516.6	1379.9

*: According to the Naval Hydrographic Office.

**: Information collected from respective crates.

As per National Hydrographic Office, Dehra Dun, the Indian sub-continent has a long coastline, extending to a length of about 7516.60 kms including Daman, Diu, Lakshadweep and Andaman and Nicobar Island. Almost all the maritime States/ UTs are facing coastal erosion problem in various magnitudes. About 1450 km of coastline has been reported to be affected by sea erosion, out of which about 700 km of coastline has been reasonably protected by construction of seawalls, groins, etc, and 750 km is yet to be protected.

State Profile

As per a survey conducted by the Ocean Engineering Division of National Institute of Oceanography, Goa, India about 23 per cent of India's mainland coastline of 5423 kms is getting affected by erosion. As much as 1248 km of the shoreline is getting eroded all along the coast with 480 km of the 569 km shoreline of Kerala (Figure 3.1) is being affected by the phenomenon. In Karnataka, the erosion was marked over 249.6 km out of the state's total coastline of 280 km with the problem relatively more severe in south Kannada and Udupi coasts where about 28 per cent of the total stretch is critical. Again, the coastal erosion is marked over 263 km of the 652.6 km shoreline of Maharashtra (Figure 3.2) while 107.6 km out of 476.4 km of Orissa's coastline has already been affected. Shoreline erosion in the northern regions of Chennai, Ennore, Visakhapatnam and Paradip ports has resulted due to construction of breakwaters of the respective port.

Rivers have been identified as the major sources of sediments along the Indian coast among which the Ganges and Brahmaputra contributed a major share of suspended sediments to the Bay of Bengal and the Indus to the Arabian Sea. The continental shelf along the country's east coast is narrow whereas along the west coast, the width of the shelf varies from about 340 km in the north to less than about 60 km in the south.

The west coast of India experiences high wave activity during the southwest monsoon with relatively calm sea conditions prevailing during the rest of the year. On the other hand, in the east coast, wave activity is significant both during southwest

Figure 3.1: Erosion at Kappakkal Beach (Payyanakkal) At Calicut, Kerala.

Figure 3.2: Erosion of Versov Coast, Maharashtra where houses washed out.

and northeast monsoons. Extreme wave conditions occurred under severe tropical cyclones which are frequent in the Bay of Bengal during the northeast monsoon period.

Along Gujarat coast, shoreline erosion is observed at Ghoga, Bhagwa, Dumas, Kaniar, Kolak and Umbergaon, and sediment deposition leading to the formation of sand spits at the estuarine mouths of the Tapti, Narmada, Dhadar,Mahe, Sabarmati, Kim, Purna and Ambika.

Erosion has been observed at Versoa, Mumbai; near Kelva fishing port, north of Mumbai and at Rajapuri, Vashi and Malvan along the **Maharashtra coast.** Along **Goa coast**, erosion is noticed at Anjuna, Talpona and Betalbatim.

The **Andaman and Nicobar** consists of about 265 islands, most of which are composed of rocks like fossiliferous marine petroliferous beds, conglomerates, sandstone and limestone. Land subsidence of 0.8 metre to 1.3 metre has occurred at the Andaman and Nicobar islands due to December 26, 2004 tsunami and has resulted in shoreline erosion in some of the islands.

The **Lakshadweep,** is an archipelago of coral islands in the Arabian sea consists of 36 islands, 12 atolls, three reefs and five submerged coral banks. Coastal erosion in all these islands had been taken up in all the inhabited islands and a total length of 40 km had been protected so far, the report said.

Erosion along the beaches near river mouths has been commonly noticed along **Karnataka coast**. Coastal erosion and submergence of land have been reported at Ankola, Bhatkal, Malpe, Mulur, Mangalore, Honnavar, Maravante and Gokarn in Karnataka. About 60 km of beach (19 per cent of the total lengthof shoreline) is affected by erosion. The problem is relatively more severe in Dakshina Kannada and Udupi coasts, where about 28 per cent of the total stretch is critical. In Uttara Kannada region, about 8 per cent of the coast is subjected to severe erosion.

Along **Tamil Nadu** coast, the erosion rate observed at Poompuhar, Tarangampadi, Nagapattinam, Mandapam, Manapadu, Ovari, Kanyakumari, Pallam,

Table 3.2: Damages due to Sea Erosion (As on 2007).

States	Land Lost		Residential/Industrial/Office Buildings		Crops		Other Losses (Plantation, public utilities, cattle, village road, boats, etc.)		Total Annual Losses (Rs. Crores)
	Quantity (Ha.)	Amount (Rs. Crores)	Quantity (No.)	Amount (Rs. Crores)	Quantity (Ha.)	Amount (Rs. Crores)	Quantity (Ha.)	Amount (Rs. Crores)	
Karnataka	67.48	3.485	2294	25.882	—	0.3109	—	1.602	31.28
Maharashtra	66.20	12.85	34	0.363	10.73	0.552	—	30.258	44.023
Odisha	89.55	2.24	3000	75	—	49.00	—	17.50	143.74
Tamil Nadu	67.55	46.453	209	6.162	271	0.906	—	7.163	60.684
West Bengal	102	12.50	28.373	9.07	74124	23.38	—	34.11	79.06
Pondicherry	—	—	—	—	—	—	—	—	9.60
Total	**392.78**	**77.528**	**33910**	**116.477**	**74405.7**	**74.149**	**—**	**90.633**	**368.387**

Manavalakurichi and Kolachel is about 0.15, 0.65, 1.8, 0.11, 0.25, 1.1,0.86, 1.74, 0.60 and 1.2 m/yr respectively. The maximum rate of erosion along Tamil Nadu coast is about 6.6 m/yr near Royapuram, between Chennai and Ennore port10. The accretion rate at Cuddalore, Point Calimere, Ammapattinam, Kilakarai, Rameswaram, Tiruchendur, Manakudi and Muttam is observed to be about 2.98, 3.4, 0.72, 0.29,0.06, 0.33, 0.57 and 0.17 m/yr respectively.

Andhra Pradesh coast has frequently been affected by cyclones and inundated by storm surges. Erosion is noticed at Uppada, Visakhapatnam and Bhimunipatnam. Erosion is noticed at Gopalpur, Paradip and Satbhaya in **Orissa**. Growth of long sand spits at Chilka lake indicates the movement of littoral sediment and subsequent deposition.

Major length of **the West Bengal** coast is represented by the Sundarban region of the Ganges mouth with shoals,sand spits, mud flats and tidal swamps. Beach erosion is noticed at Digha, Bankiput and Gangasagar regions of the West Bengal coast. In Kerala, about 360 km long coastline is exposed to erosion.

Preventive and Mitigation Measures

Structural Measures

☆ Through construction of Seawall/Revetment

☆ Groynes

☆ Offshore breakwater

Non-Structural/Soft Measures

☆ Artificial nourishment of beaches

☆ Vegetation cover

☆ Sand bypassing at tidal inlets

The remedial measures should be selected after proper investigation and model studies. It must be ensured that protection measures do not shift erosion problem from one site to some other site. The measures to control erosion include non-structural and structural or their combination.

Structural measures used for coastal erosion prevention are as follows:

☆ *Sea wall/Revetment*: Seawall may be useful in case of protection of specific area from erosion and storm surges. Adverse effect is experienced on downstream side.

☆ *Groynes*: Groynes may be adopted to stop or decrease shoreline recession and for beach formation. However, extremely adverse effects are observed on downstream side and groynes should be avoided unless their main purpose is to keep a beach at one particular position at the cost of adjoining areas.

☆ *Off-shore breakwater* : These may be adopted for shore protection and beach formation. Severe downstream erosion may result due to littoral barrier effect. It is an expensive option and needs regular maintenance to avoid rapid breakdown of breakwater.

Soft-structural measures generally adopted to reduce/prevent coastal erosion are:

☆ *Artificial nourishment of beaches* : Beach nourishment may be adopted for protection and beach development. Combination of nourishment of beaches with seawall/groynes will create beach in front of protected area and eliminate leeside erosion.

☆ *Vegetation cover such as mangrove and Palm plantation* : Vegetation cover can restrict sand movement and erosion.

☆ *Sand bypassing at tidal inlets* : Severe erosion problem has been experienced due to construction of jetties and/or dredged channels. This problem can be solved by bypassing of material from the updrift side of inlet to the downdrift side.

Out of these measures, the techno-economically viable and site-specific suitable measure should be adopted. Combination of the above measures may give optimum results with least adverse effect on down drift.

Use of Gryones

Appropriate Locations

High value frontages influenced by strong long shore processes (wave induced or tidal currents) where nourishment or recycling are undertaken. Best on shingle beaches or within estuaries.

Effectiveness

Good on exposed shorelines with a natural shingle upper beach. Can also be useful in estuaries to deflect flows. Unlimited structure life for rock groynes.

Benefits

Encourages upper beach stability and reduces maintenance commitment for recycling or nourishment.

Problems

Disrupts natural processes and public access along upper beach. Likely to cause downdrift erosion if beach is not managed.

General Description

Groynes are cross-shore structures designed to reduce longshore transport on open beaches or to deflect nearshore currents within an estuary. On an open beach they are normally built as a series to influence a long section of shoreline that has been nourished or is managed by recycling. In an estuary they may be single structures.

Rock is often favoured as the construction material, but timber or gabions can be used for temporary structures of varying life expectancies (timber: 10-25 years, gabions: 1-5 years). Groynes are often used in combination with revetments to provide a high level of erosion protection.

Figure 3.3: Recently built rock groyne at estuary mouth, constructed in association with beach renourishment of adjacent foreshore.

Function

Groynes reduce longshore transport by trapping beach material and causing the beach orientation to change relative to the dominant wave directions. They mainly influence bedload transport and are most effective on shingle or gravel beaches. Sand is carried in temporary suspension during higher energy wave or current conditions and will therefore tend to be carried over or around any cross-shore structures. Groynes can also be used successfully in estuaries to alter nearshore tidal flow patterns. Rock groynes have the advantages of simple construction, long-term durability and ability to absorb some wave energy due to their semi-permeable nature. Wooden groynes are less durable and tend to reflect, rather than absorb energy. Gabions can be useful as temporary groynes but have a short life expectancy. Groynes along a duned beach must have at least a short "T" section of revetment at their landward end to prevent outflanking during storm events. The revetment will be less obtrusive if it is normally buried by the foredunes. Beach recycling or nourishment is normally required to maximise the effectiveness of groynes. On their own, they will cause downdrift erosion as beach material is held within the groyne bays.

Beach Drainage

Appropriate Locations

Low tidal range sand beach sites with a high amenity value, low to moderate wave energy

Effectiveness

Increases upper beach width and therefore dune stability, variable life expectancy.

Benefits

Non-intrusive technique resulting in wider, drier beach

Problems

Storm erosion of beach is likely to damage the system

General Description

Beach drains comprise perforated land drain pipes buried below the upper beach surface, and connected to a pump and discharge. The concept is based on the principle that sand will tend to accrete if the beach surface is permeable due to an artificially lowered water table. The system is largely buried and therefore has no visual impact.

Function

Mild upper beach and dune erosion can be controlled by beach drains. The system actively lowers the water table in the swash zone, thereby enhancing the wave absorption capacity of the beach, reducing sand fluidisation and encouraging sand deposition. The deposited sand forms an upper beach berm that protects the dune face during storm events that might otherwise cause erosion.

Benefits are greatest on micro-tidal (<2m range), high value amenity beaches where landscape issues preclude the use of other management approaches. Important backshore assets should not rely on drainage systems for erosion protection during storms, even as a temporary measure.

Rock Revetments

Appropriate Locations

Sites suffering severe and ongoing erosion where important and extensive backshore assets are at risk.

Effectiveness

Good long-term protection. Can be extended or modified to allow for future shoreline change. Unlimited structure life.

Benefits

Low risk option for important backshore assets. Permeable face absorbs wave energy and encourages upper beach stability.

Problems

Strong landscape impact. Can alter dune system permanently as sand tends not to build up over the rocks if beach erosion continues.

General Description

Rock revetments may be used to control erosion by armouring the dune face. They dissipate the energy of storm waves and prevent further recession of the backshore if well designed and maintained. Revetments may be carefully engineered structures

Figre 3.4: Major rock armour revetment in front of dune system.

protecting long lengths of shoreline, or roughly placed rip-rap protecting short sections of severely eroded dunes.

Though offering long-term security, the landscape impact and damage to habitat are considerable.

Function

Rock revetments are widely used in areas with important backshore assets subject to severe and ongoing erosion where it is not cost effective or environmentally acceptable to provide full protection using seawalls he function of permeable revetments is to reduce the erosive power of the waves by means of wave energy dissipation in the interstices of the revetment.

Permeable revetments can also be built from gabions timber) or concrete armour units. Concrete units are normally too costly for use as dune protection, but may be appropriate where high value back shore assets must be protected and armour rock is difficult to obtain. They are often considered to be more unattractive than rock.

Revetments may not prevent on going shoreline recession unless they are maintained, and, if necessary, extended. If the foreshore continues to erode, the rock revetment may slump down, becoming less effective as a defence structure, but will not fail completely. Repairs and extensions may be necessary to provide continued backshore protection at the design standard.

Impermeable Revetments and Seawalls

Appropriate Locations

Exposed frontages with extensive and high value backshore assets.

Effectiveness

Provides good medium term protection, but continued erosion will cause long term failure (30-50 year life expectancy).

Benefits

Fixed line of defences allowing development up to shoreline. Allows amenity facilities along backshore and easy access to beach.

Problems

Continued erosion may cause undermining and structural failure. Complete disruption of natural beach-dune processes.

General Description

Impermeable revetments are continuous sloping defence structures of concrete or stone blockwork, asphalt or mass concrete. Revetments are built along the dune face, preferably above the run-up limit of waves under normal conditions. Where frequent wave attack is anticipated, the revetment may be topped by a vertical or recurved wall to reduce overtopping. Seawalls are near vertical structures of concrete, masonry or sheet piles, designed to withstand severe wave attack. Their use was popular in the past but they are now normally considered to be costly, detrimental to the stability of beaches and unsuitable for erosion management along a dune shoreline.

Function

The rock armour was placed after beach lowering exposed the toe of the revetment. The good medium term protection of such structures has to be balanced against considerable landscape impact and habitat damage.

Figure 3.5: Rock faced concrete revetment with sheet piled toe and rock armour apron.

Figure 3.6: Concrete and stone seawall.

Impermeable revetments provide a fixed line of defence for frontages with high value backshore assets. They are intended to withstand storm wave attack over a life expectancy of 30 to 50 years. Amenity facilities such as promenades, slipways and beach access steps can be built into the revetment.

Revetments will severely disrupt natural beach-dune interactions, and should not be used on frontages valued for natural heritage. Ongoing beach erosion may result in undermining of the revetment toe, leading eventually to structural failure or the need for repairs and extensions.

In common with many other fixed structures the natural interchange of sand between beach and dunes is prevented with the consequent loss of transitional habitats.

Artificial Seaweed

There have been several attempts at placing artificial seaweed mats in the nearshore zone in an effort to decrease wave energy by the process of frictional drag. The field trials have generally been inconclusive as regards wave energy attenuation. The most successful trials have been in areas of very low wave conditions, low tide range and relatively constant tidal current flows, when some sedimentation was found to take place.

On open coast sites there have been major problems with the installation of the systems and the synthetic seaweed fronds have shown very little durability even under modest wave attack. The synthetic seaweed has tended to flatten under wave action, thereby having minimal impact upon waves approaching the coast.

Field trials in the United Kingdom have been unsuccessful and the experiments were abandoned in all cases, due to the material being ripped away from the anchorage points.

In the Netherlands experiments were more successful with synthetic seaweed being placed in relatively deep water, where sedimentation up to 0.35m took place soon after installation, although this would result in only a very minor decrease in shoreline wave conditions.

The cost of the artificial seaweed is low but the costs and frequency of maintenance works make this option not worth pursuing in an exposed coastal environment, where it would be subject to severe wave conditions and would become damaged rapidly.

Alternative Breakwaters

A considerable amount of research has been carried out on the potential performance of various types of breakwater including:

☆ Layered plate frameworks

☆ Floating breakwaters

☆ Perforated caissons

These techniques involve the attenuation of wave energy by means other than providing a direct barrier. The numerous designs that have been tested or built are usually specific to a particular wave environment, and are usually aimed at vessel protection over relatively short distance. Design, construction and management costs are high. None have been shown to be practical as far as dune protection is concerned.

Other Comments

Literature on the CRZ Notification is easily available in India in printed form. An excellent compilation is available from The Goa Foundation, Above Mapusa Clinic, Mapusa 403 507, Goa, India. The book contains the CRZ Notification, the Supreme Court judgement upholding it and the approved CZMPs of all the coastal states in India.

References

Ministry of Environment and Forests, Govt. of India: www.envfor.nic.in

Coastal Zone Management Handbook: John R. Clark, Lewis Publishers, 1996

Report of the committee chaired by prof. M.S. Swaminathan to review the coastal regulation zone notification, 1991 (February, 2005)

Coastal Zone Management (CZM) Notification 2007, Ministry of Environment and Forests

Citation: UNEP-WCMC; In front Line: Shoreline Protection and other ecosystem services from mangroves and coral reefs, UNEP-WCMC, Cambridge, UK, 2006

Vulnerability of Indian Coastline to sea level rise; Diksha Aggarwal and Murari Lal: Centre for Atmospheric Sciences, Indian Institute of Technology, New Delhi, India

India's policy for protecting the coastal environment for sustainable use; Environmental policies and management, Ministry of environment and forests, Govt. of India, New Delhi

Presentation by A.K.Kharya, Central Water Commission, New Delhi on "Basic concepts and effects of coastal erosion" on 18 September 2006 at National Institute of Disaster Management, New Delhi.

Presentation by Dr. A. Senthil Vel, Additonal Director, Ministry of Environment and Forests, Govt. of India, New Delhi on "Regulatory Framework for Coastal Zone Management and Coastal Erosion" on 18 September 2006 at National Institute of Disaster Management, New Delhi.

Presentation by C.K.L. Das, Director (coastal erosion), Central Water Commission, New Delhi on "Coastal Erosion Management: Recent Initiatives" on 20 september 2006 at National Institute of Disaster Management, New Delhi.

4

Bioterrorism: A Technological Hazard for Ecology and Mankind

Manish Kumar Verma, Brijesh Kumar Singh, Rajeev Niranjan and Anil K. Gupta

Introduction

Biological and chemical weapons have had a long and checked history. The use of biological weapons and efforts to make them for more efficient for waging wars has been repeated numerous times. Threat of bioterrorism has long been ignored and has been denied proper consideration in civil administration. However it has sought and attained significance attention during past few years. Recent tragidic events in Iraq, Japan and Russia cast an ominous shadow.

Bioterrorism may occur as covert events, in which persons are unknowingly exposed and an outbreak is suspected only upon recognition of unusual disease, clusters and symptoms. For potential bioterrorism related condition that are endemic and low incidence, the use of nontraditional surveillance methods and complementary data resources may enhance our ability to rapidly detect changes in disease incidence. In response to global bioterrorism threats, CDC (Center for Disease Control and Prevention) has proposed a list of critical biological agents that have potential for use in a terrorist incidence.

Present review makes an overview of causative agents for means of bioterrorism and discusses the mode of risk imposed. Issues related to prevention of bioterrorism on short term objectives and also for long term strategies have been discussed.

Keywords: Biological weapons, Disease clusters and Symptoms, Prevention.

The FBI defines terrorism as *"the unlawful use of force or violence against persons or property to intimidate or coerce a Government, the civilian population, or any segment thereof, in furtherance of political or social objectives."* (Chandler and Landrigen, 2004; FBI, 2010; DOD, 2009).

Today, terrorists increasingly aim for mass casualties, panic and death. In the years since that attack, while there have been no further mass incidents on U.S. soil, other large-scale terrorist attacks have taken place in Indonesia (on the island of Bali in October 2002), Iraq (especially after the war was declared over by U.S. officials), and Spain (in Madrid in March 2004). The advent of mass casualty terrorism-and the reports of some terrorist organizations' pursuit of unconventional chemical (Use of non-explosive chemical agents that are not themselves living organisms, to cause injury or death. (Encyclopedia, The freedictionary.com, Chemical warfare, 2010), biological (Use of any organism, like bacteria, virus or other disease-causing organism, or toxin found in nature, as a weapon of war. It is meant to incapacitate or kill an adversary. (Encyclopedia, The freedictionary.com, Biological warfare, 2010), radiological (A radiological weapon or radiological dispersion device, RDD) is any weapon that is designed to spread radioactive contamination, either to kill, or to deny the use of an area (a modern version of salting the earth) and consists of an device (such as a nuclear or conventional explosive) which spreads radioactive material. They have recently been called "dirty bombs. (Encyclopedia, The freedictionary.com, Radiological weapon, 2010) and nuclear weapons (A nuclear weapon is a weapon that derives its energy from nuclear reactions and has enormous destructive power - a single nuclear weapon is capable of destroying a city.(Encyclopedia, The freedictionary.com, Nuclear weapon, 2010) - indicates that the world is seeing a new type of terrorism altogether. This new terrorism could use anything from salmonella and smallpox to dirty bombs and hijacked airplanes as weapons of mass destruction. Terrorism is not just about inflicting harm or damage; it is about instilling fear. Even hoaxes and threats can terrorize large populations, causing social and economic harm even when no real danger exists. (Chandler and Landrigen, 2004).

During past nine years, the threat of bioterrorism has become a subject of widespread concern. Journalists, academics, and policy analysts have considered the subject, and in most cases found much to alarm them. Most significantly, it has captured the attention of policy makers at all levels of government in the United States. Unfortunately, bioterrorism remains a poorly understood subject. There is no commonly accepted definition of bioterrorism. For purposes of this study, *"Bioterrorism is assumed to involve the threat or use of biological agents (or biological weapons) by individuals or groups motivated by political, religious, ecological, or other ideological objectives"* (FBI, 2010). Biological weapons could be just as destructive as chemical and nuclear weapons, but they are all the more frightening because they strike silently, invisibly, and may not even be discovered until long after the attack, giving the attackers plenty of time to flee far from the scene.

Bioterrorism can range from putting waste matter into food in a small-town restaurant to the aerosolized release (dispersing an agent in a particulate form) of a contagious virus over a large city, or even the spreading of plant or animal diseases in farming areas to disrupt the nation's food supply. The perpetrator can be anyone from a disgruntled employee to a hostile foreign nation or transnational terrorist group. The type of biological agent used, the means of dissemination, and the effectiveness of the response, as well as unpredictable variables such as rainfall and wind, will determine how many people are affected over how wide an area, and how severe their symptoms are. Theoretically, the number of potential biological agents is almost limitless, but certain agents naturally have a combination of properties (such as hardiness, transmissibility and virulence) that make them most effective as weapons. Several of these have been developed and tested for use as biological weapons, and these are the ones considered most likely to be used in a terrorist attack (Chandler and Landrigan, 2004).

Biological vs. Chemical and Nuclear Weapons

Figure 4.1: Symbols: Nuclear, biological and chemical weapons.

The differences between nuclear or chemical attacks and biological attacks can be compared to the difference between air strikes and sabotage missions. While both are methods of attack, they are based on different technologies, unfold differently and have vastly different effects. Aside from usually being detectable by smell and sometimes by sight (as in the greenish-yellow color of chlorine gas), chemical agents work by creating relatively immediate physical effects in those exposed-usually via the skin, respiratory system, digestive system and/or neurological system. Decontamination usually attenuates the symptoms, and while high levels of exposure may have fatal or lingering effects, the attack is over as soon as the chemical no longer is being disseminated. The immediate and finite aspects of chemical weapons make them comparable to an air strike; the attack has a noticeable beginning, it inflicts damage and ends quickly, and it allows damage assessment and consequence management to begin almost immediately. Nuclear weapons, even more so than chemical, produce a dramatically obvious initial blast that causes immediate damage in a clearly defined area. Unlike chemical weapons, however, nuclear contamination can also leave a lasting legacy of latent, invisible cancers and mutations that may take decades to develop.

In the case of biological weapons, the crisis is measured in weeks and months, not minutes and hours. Even the fact that an act of biological terrorism had taken place could, and probably would, escape detection for days or weeks because detection currently depends on public health systems' ability to recognize unusual infections or upsurges in reported symptoms-symptoms that initially might resemble nothing more serious than the flu. In this sense, a biological attack is more like an undercover sabotage mission-the destructive blow is not immediately apparent and only time will reveal the attack's nature and extent. Whereas chemical or nuclear weapons attacks would be followed by a large, immediate response by federal response teams and/or local fire departments and emergency medical services, biological attacks would produce a delayed response requiring difficult coordination among local hospitals, state and local public health departments, and the federal public health system. Chemical weapons are often easier and cheaper to produce and easier to deploy than biological weapons. Chemical weapons often are closely analogous to industrial-use chemicals (*e.g.*, the nerve agents Sarin and Tabun are closely related to industrial pesticides). Therefore, terrorists can steal industrial chemicals to use as weapons, and due to the commercial use of these chemicals, the technology to manufacture them is relatively widely circulated and relatively easily copied. By contrast, the technology, materials and expertise required to develop nuclear weapons are by far the most expensive and difficult of all the so called "weapons of mass destruction." For example, one U.S. government study concluded that nuclear weapons would cost $1,500 per person killed, while anthrax could cause deaths at a cost of just a penny each. The contrast is so great that biological weapons have been referred to as the "poor man's nuclear weapon." Some studies estimate that anthrax spores, correctly prepared, could be 1,000 times more lethal and could infect an area 1,000 times larger than the same weight of Sarin, one of the more potent chemical nerve agents. In addition to potentially extreme physical harm, the most widespread damage caused by a biological agent may well be psychological. In some cases, there will be no clearly defined specific area to fear and avoid, so instead people may develop a generalized fear of public places, going outdoors, opening the mail or even breathing. (Chandler and Landrigan, 2004; Encyclopedia, The freedictionary.com, Weapons of mass destruction, 2010; Mayer, 1995)

History

The concept of biological agents as weapons is hardly a novel idea. History offers examples, tempered by the existing levels of scientific knowledge about infectious diseases, of the use of biological agents for inflicting harm upon enemies. Long before the germ theory of disease was advanced, humans associated disease with foul odors; contagion was thought to be spread by "miasmas," or bad vapors (Eickhoff, 1996). Biological warfare has been practised repeatedly throughout history. In 184BC, Carthaginian leader Hannibal had clay pots filled with poisonous snakes and instructed his soldiers to throw the pots onto the decks of Pergamene ships (Encyclopedia, The freedictionary.com, Biological warfare, 2010).

In the first documented cases of biological warfare in the 1340s, Europeans catapulted dead bodies into besieged cities and castles in the hope of causing unlivable conditions and spreading infections such as plague. By the 1420s, they had added animal manure to increase infections caused by the rotting cadavers (Chandler and Landrigan, 2004). The concept of inanimate fomites as vehicles for spreading disease to enemies was chronicled in the 18th century. In 1763, Sir Jeffrey Amherst, the commander of British troops in North America, was concerned about activities of Native Americans along the western frontier (extending from Pennsylvania to Detroit) who were unsympathetic to the British. When he learned that smallpox had broken out among British troops at Fort Pitt, he suggested that the disease could be used as a biological weapon against the Native Americans. The plan was to pass along blankets or handkerchiefs used by the British smallpox victims to the hostile Native Americans. An epidemic of smallpox did occur among these Native American tribes, but it cannot be assumed that the outbreak resulted from biological warfare activities of the British. The Native Americans were immunologically naive as far as smallpox was concerned and had many opportunities to contract the disease in other contacts with European settlers. Historical evidence also suggests that the French used smallpox as a weapon against Native Americans during this era (Christopher *et al.*, 1997; Poupard and Miller, 1992). All of the above activities, which may be considered early attempts at biological warfare, occurred before the germ theory of disease was formulated and widely accepted (Christopher *et al.*, 1997). There was some evidence of biological warfare in World War I. During World War I, Germany was thought to have employed the agents of cholera and plague against humans and anthrax and glanders against livestock (Christopher *et al.*, 1997; Poupard and Miller, 1992).

In the period between World Wars I and II, a number of countries, including the USSR, Japan, and the United Kingdom, stepped up their biological warfare research programs. The Japanese effort was notable, with a number of military units engaged in offensive biological weapons research until the end of World War II. One of the most notorious of these, Unit 731, was headed by Army physician-microbiologist Ishii Shiro from its inception in 1937 until 1941. Unit 731, the second such unit that Ishii had commanded, was located in Japanese-occupied Manchuria. At the height of its operations, the unit's staff of 3,000 was quartered in 150 buildings at Ping Fan. Unit 731 personnel also oversaw at least five satellite operations, each with own staff of 300 to 500. This biological warfare unit and others like it were responsible for extensive research and development, using both animal and imprisoned human subjects (usually criminals or political dissidents). It is estimated that during 13 years of Japanese biological warfare research in Manchuria and China, 10,000 unwilling human "subjects" lost their lives. An extensive menu of bacterial, viral, and rickettsial diseases was investigated during the Japanese effort in the 1930s and early 1940s. The Japanese also conducted at least a dozen field tests in Manchuria and China. These tests included the contamination of water and food supplies, aerial spraying, and the dropping of small bombs containing plague-infected fleas. Outbreaks of plague, cholera, and typhus were attributed to these activities (Harris, 1992). Biological weapons during World War II, there are no evidence to indicate they were actually used on a large scale. There is, however, strong evidence that Reinhard Heydrich, chief of the Nazi security service, was assassinated with a grenade that had been contaminated with biological warfare agents (typhoid fever) (Mayer, 1995).

Biological Warfare in the Cold War

After World War II and during the Korean War, the focus, at least from the United States perspective, was on building a biological warfare retaliatory capability.

The US developed an anticrop bomb and delivered it to the Air Force in 1951. It could have been used to attack North Korean rice fields, reducing a significant source of nutrition for the population (Mayer, 1995). North Korea accused the United States of using biological agents during the Korean War; the United States denied the accusation, and there was no substantive proof offered in the open literature (Harris, 2002).

Following the Korean War, the United States invigorated the biological warfare program in 1956 after Marshal Zhukov announced to the Soviet Congress that chemical and biological warfare weapons would be used as weapons of mass destruction in future wars. This was a dramatic shift in Soviet policy and the cold war philosophy (US Army, 1997). The fundamental concept of United States biological warfare operations changed as a result.

In the latter half of the 20th century[1984], the only event confirmed as a successful act of bioterrorism was the one carried out by the followers of Baghwan Shree Rajneesh. In a small town of The Dalles in Oregon, followers of the Bhagwan Shri Rajneesh (the Rajneeshee Cult) attempted to control a local election by infecting salad bars with *Salmonella*. The attack caused about 900 people to get sick. It is considered the first ever bioterrorism case in the US history (Encyclopedia, The freedictionary.com, Biological warfare, 2010; Chandler and Landrigan, 2004).

Gulf war (1991), By the time of the Iraqi invasion into Kuwait, it was widely acknowledged that Iraq had a biological warfare program, concentrated on very toxic *botulinum* toxin and very resilient anthrax (DOD, 1992). This assessment was derived from a compilation of several sources and indicators, the most dramatic being an Iraqi defector who was a microbiologist. He told a British newspaper correspondent that as early as 1983 Iraqi scientists were developing and testing biological warfare agents: There were many strains, botulism, *salmonella*, and anthrax (Tucker, 1993).

Saddam Hussein announced "loud and clear" that this war would be the "mother of all wars," implying a no-holds-barred engagement. This was the first time since World War II that the United States had faced a military adversary with a highly probable biological warfare capability and the resolve to use it (Mayer, 1995).

India's defense and intelligence outfits were alert to the outbreak of pneumonic plague-well known in biological warfare-in Surat and bubonic plague in Beed in 1994, which caused several deaths and sizeable economic loss. "The *Yersinia pestis* strains that are percolating in established plague foci in India are very much less virulent and definitely different from the samples we have seen from the plague outbreak region," said Dr Harsh Batra, joint director of the Defense Research and Development Establishment, who led the studies on outbreaks of plague. In the absence of more data and samples, he refused to attribute the 1994 plague conclusively to external biowarfare (Sherma, 2001).

The anthrax attacks through the U.S. mail in October 2001 demonstrated that, with access to a highly refined agent, a damaging through bioterrorism can be delivered with only an envelope and a stamp. The anthrax attacks were the first terrorist biological attacks to garner immediate worldwide attention. While the investigation continues and many questions remain, it is known that the anthrax used was highly refined, highly lethal and probably originated in a sophisticated laboratory, most likely a U.S. government lab. Another recent event has underscored the uncertainties that can accompany a possible biological attack. When the new disease now known as SARS (Severe Acute Respiratory Syndrome) was first detected early in 2003, there were some initial fears that it might be a biological toxin being deliberately spread through the world's population. It took several weeks of study to determine its natural origins and pattern of spread, apparently from initial infections in Southeast Asia, and to identify the virus responsible. But it is entirely possible that some future attack could unfold in a similar way, leaving people-including doctors and public health officials-uncertain for an extended period about whether they were facing a natural epidemic or a deliberate act of terrorism (Chandler and Landrigan, 2004).

Humans are not the only species targeted by biological weapons and increased pathogenic virulence. From crows and eagles in North America, to lions in Africa, to frogs and amphibians around the world, environmental stresses are taking a heavy toll. The tempo of this problem is increasing rapidly, as increased environmental pressures affect both humanity as well as other organisms. One example of this is the unexplained spread of Phocine Distemper virus (PDV) amongst seal populations around the world. PDV is closely related to canine distemper and has apparently always afflicted seal populations to one degree or another. However, in recent years there have been a series of increasingly lethal PDV epidemics amongst seal populations including one in 1998 that wiped out half of Europe's seals. An even more amazing instance is the massive global decline in amphibian populations. Around the world - even in relatively pristine areas- amphibian populations have been crashing. Frogs are particularly hard-hit. In some areas entire species have vanished without a trace in the space of just a few years. The causes appear to be quite complex and remain to be clarified, but the spread of increasingly pathogenic viral and fungal infections seem to be a key element. Amphibians have fragile physiologies and therefore are extremely sensitive to environmental disruption. Another recent example is the introduction of the West Nile virus into a new ecology - North America. As its name implies, West Nile originated in tropical Africa. Apparently it recently jumped to the United States, found the warming climate to it's liking, and proceeded to sicken and kill a number of people. However, it's far more deadly to other species. In some situations it appears to kill 100 per cent of infected American Crows. Many other bird species, in particular raptors and owls, are also extremely vulnerable. Yet another example is the recent cancer epidemic afflicting Australia's Tasmanian Devils. Tasmanian Devils are a predatory marsupial confined to - you guessed it - Tasmania. In prehistoric times they ranged in mainland Australia as well, but were apparently eliminated by the aboriginal introduction of the dingo. In any event, the devils are now faced by a retrovirus which induces a fast-growing and fatal form of cancer. The species will probably survive this onslaught, even though it looks like the majority of the population might eventually succumb.

The plant world is just as vulnerable to such epidemics and the results are evident. For instance, in the western United States oak trees are rapidly dying due to a new disease called, appropriately enough, Sudden Oak Death. This disease is caused by the previously unknown fungus *Phytophthora ramorum*, which is genetically closely related to the species that caused the great Irish potato famine (another classic example of a plant epidemic, by the way). *Phytophthora* infects over 20 species of plants, including redwoods, although it appears to be particularly deadly in three of these species. Again, no one knows why this fungus has appeared and is exacting such a huge cost. But, as in the previous examples, it is likely that environmental disruption is the key catalyst (Zkea Archives, 2010).

Potential of Biological Weapons

Biological weapons pose the greatest threat as these are ideal for bio terrorism for the following reasons (Debashis, 2001).

☆ Have a delayed response thus preventing immediate detection

☆ The easiest to acquire

☆ Less expensive

☆ Not easily detected

☆ Even a small quantity can be fatal affecting masses

☆ Has potential for major public health impact

☆ Might cause public panic and social disruption

☆ Require special action for public health preparedness.

For example, anthrax is considered an excellent agent. We use it here because it is historically important and enough information is public that this discussion can't be a manual. First, it forms hardy spores, perfect for dispersal aerosols. Second, pneumonic (lung) infections of anthrax usually do not cause secondary infections in other people. Thus, the effect of the agent is usually confined to the target. A pneumonic anthrax infection starts with ordinary "cold" symptoms and quickly becomes lethal. Finally, friendly personnel can be protected with suitable antibiotics or vaccines.

A mass attack using anthrax would require the creation of aerosol particles of 1.5 to 5 micrometres. Too large and the aerosol would be filtered out by the respiratory system. Too small and the aerosol would be inhaled and exhaled. Also, at this size, nonconductive powders tend to clump and cling because of electrostatic charges. This hinders dispersion. So, the material must be treated with silica to insulate and discharge the charges. The aerosol must be delivered so that rain and sun does not rot it, and yet the human lung can be infected. There are other technological difficulties as well. Diseases considered for weaponization, or known to be weaponized include anthrax, Ebola, pneumonic plague, cholera, tularemia, brucellosis, Q fever, Machupo, VEE, SEB and smallpox. Naturally-occurring toxins that might be used in weapons include ricin, botulism toxin, and mycotoxins (Encyclopedia, The freedictionary.com, Biological warfare, 2010). Biological Agents are grouped under three categories based on the potency and ease of dispersion. Anthrax comes under the top category for its powerful effect and easy availability. One redeeming feature is that it can be cured with antibiotics if detected in time.

Category A

High-priority agents include organisms that pose a risk to national security because they can be easily disseminated or transmitted person-to-person; cause high mortality, with potential for major public health impact; might cause public panic and social disruption; and require special action for public health preparedness.

Agents include *variola major* (smallpox); *Bacillus anthracis* (anthrax); *Yersinia pestis* (plague); *Clostridium botulinum* toxin (botulism); *Francisella tularensis* (tularaemia); filoviruses; Ebola hemorrhagic fever, Marburg hemorrhagic fever, and arenaviruses; Lassa (Lassa fever), Junin (Argentine hemorrhagic fever) and related viruses.

Category B

The second highest priority agents include those that are moderately easy to disseminate; cause moderate morbidity and low mortality; and require specific enhancements of CDC's diagnostic capacity and enhanced disease surveillance.

Agents are *Coxiella burnetti* (Q fever); *Brucella species* (brucellosis); *Burkholderia mallei* (glanders); alpha viruses, Venezuelan encephalomyelitis, eastern and western equine encephalomyelitis; ricin toxin from *Ricinus communis* (castor beans); epsilon toxin of *Clostridium perfringens*; and *Staphylococcus* enterotoxin B. A subset of List B agents includes pathogens that are food or waterborne. These pathogens include but are not limited to *Salmonella* species, *Shigella dysenteriae*, *Escherichia coli* O157:H7, *Vibrio cholerae*, and *Cryptosporidium parvum*.

Category C

The third priority agents include emerging pathogens that could be engineered for mass dissemination in the future because of availability; ease of production and dissemination; and potential for high morbidity and mortality and major health impact.

Category C agents include Nipah virus, hantaviruses, tickborne hemorrhagic fever viruses, tickborne encephalitis viruses, yellow fever and multi drug resistant tuberculosis. (CDC.3, 1998; Debashis, 2001).

Source of Biological Agents

Biological agents are organisms or toxins produced by organisms that can be used against people, animals, or crops. In contrast, chemical agents, poisonous substances that can kill or incapacitate, are manmade materials (Sidell, 1997; Eitzen, 1997).

Pathogens

Pathogens are naturally occurring microorganisms that cause disease. There are hundreds of pathogens, including bacteria, viruses, fungi, and parasites. Among the pathogens often mentioned as potential biological agents are *Bacillus anthracis*, the organism that causes anthrax, and *Yersinia pestis*, the organism that causes plague. Because pathogens are living organisms, they are self replicating. Exposure to even a small number of organisms can produce severe symptoms or even death. Thus, it is believed that the ID50 for pneumonic plague is fewer than 100 *Y. pestis* organisms, while 8 10,000 *B. anthracis* spores will cause inhalation anthrax (Meselson, 1995; Meselson, 2001; Watson and Keir, 1994). Only some pathogens are transmissible from person to person. For example, someone suffering from pneumonic plague can transmit *Y. pestis* organisms to others, creating a serious risk of epidemic spread. In contrast, bubonic plague is communicable generally only if someone is exposed to pus from an infected person. Anthrax is not contagious, and only those exposed to the released *B. anthracis* spores are likely to become infected (Benenson, 1995).

Pathogens require an incubation period before symptoms of infection appear. For some diseases, the incubation period is only a few days, while for others it might be several weeks. Typically, 3-5 days pass before the acute symptoms of inhalation anthrax appear, while for Q fever (caused by the *Coxiella burnetii* organism) the incubation period is two to three weeks, depending on the size of the dose (Carus, 2001; Meselson. 1995).

Toxins

Toxins are poisonous chemicals produced by living organisms. Among the best known are botulinum toxin, which is produced by the bacteria *Clostridium botulinum*, and ricin, which is extracted from the seed of the castor bean plant. Unlike pathogens, toxins are not self-replicating, so their physical effects are solely a result of the agent released. While toxins share many characteristics with chemical agents, they also have some significant differences (Sidell *et al.*, 1997). Many toxins are more toxic than the most lethal of chemical agents. Thus, the LD50 for botulinum toxin when injected is 0.001 micrograms per kilogram of body weight. In contrast, VX, perhaps the most lethal of the chemical agents, has an LD50 of 15 micrograms per kilogram of bodyweight.28 Toxins are not volatile, unlike many chemical agents, and thus do not naturally generate a persistent threat. Generally, toxins are not dermally active, meaning that contact with the skin is insufficient to produce disease. Rather, the agent must be brought into the body, either by ingestion, inhalation, or through an opening in the skin (Carus, 2001). The quantity of toxin required to achieve a desired effect is dependent on the lethality of the agent. According to one estimate, eight tons of ricin would be needed to blanket an area to achieve the same effect accomplished using only eight kilograms of botulinum toxin. For many toxins, the quantities of agent required to produce a given effect are similar in size to that for the more lethal chemical agents (Sidell *et al.*, 1997).

Combinations

The perpetrators considered use of both pathogens and toxins. Some perpetrator was interested in both *B. anthracis* and botulinum toxin. Some perpetrators thought about HIV and tetanus in addition to both *B. anthracis* and botulinum toxin and some involved in combinations that are more unusual: tetanus and botulinum toxin and *S. typhi* and an unknown mushroom poison (Carus, 2001).

Despite efforts to restrict the illicit acquisition of biological agents, it is likely that terrorists and criminals will be able to obtain the agent that they want when they want it. If unable to acquire from a legitimate culture collection or a medical supply company, they can steal it from a laboratory. If unable to steal it, a group with the right expertise could culture the agent from samples obtained in nature.

Culture Collections

The non-state actors obtained biological agents or toxins from legitimate suppliers. The American Type Culture Collection (ATCC) was the source of the agent The Rajneeshees purchased their seed stock of *Salmonella typhimurium* from a medical supply company. Significantly, another group could do the same today under current regulations, because the organism involved is not on any control list. In any case, the Rajneeshees had a state certified clinical laboratory, which gave them a legitimate reason to acquire agents like the one that they used.

Theft

The perpetrators acquired their biological agents by stealing them from research or medical laboratories. Almost all of the thefts involved people who had legitimate access to the facilities where the biological agents were kept. In only one of the reported incidents was an attempt made to infiltrate a laboratory to steal a biological agent. It is alleged that the Weathermen group attempted to suborn an employee at the U.S. military research facility at Ft. Detrick in order to obtain pathogens.

Self-manufacture

The perpetrators manufactured the agent themselves. In every reported case, the perpetrators produced ricin toxin by extracting it from castor beans. The Minnesota Patriots Council, which produced a small quantity of ricin toxin, made it from a recipe found in a book. In contrast, there were no successful attempts to grow *C. botulinum* to produce botulinum toxin. Several readily available "how-to" manuals purport to describe techniques for producing botulinum toxin or extracting ricin from castor beans. A considerable number of perpetrators have used The Poisoner's Handbook and Silent Death'. Maynard Campbell's 'Catalogue of Silent Tools of Justice', which is apparently no longer in print, was used by the Minnesota Patriots Council in their ricin production. Long before the Internet, perpetrators found it easy to obtain information about biological agents.

Natural

The biological agent was obtained from a natural reservoir and transmitted without any processing. Some perpetrators apparently used castor beans, the seed containing ricin, to poison people without attempting to extract the ricin from the bean. Similarly, in several cases the perpetrators injected the victims with HIV-contaminated blood (Carus, 2001).

Way of Use of Biological Agents

A Biological Agent is not Necessarily a Biological Weapon

Only if there is a mechanism for spreading the agent is it transformed into a weapon. Thus, a pathogen growing on a petri dish is not a weapon, or even a threat, because it is unlikely to infect anyone. If the agent is highly contagious, infecting a single person or animal may be sufficient to start an epidemic. When the agent is not contagious, as with many pathogens and all toxins, it is necessary to have a dissemination mechanism that spreads the agent to the intended target. While it is possible to infect people by injecting them one by one with biological agents, such a method is unlikely to prove attractive to most terrorists. More likely, a terrorist will seek a technique to infect the entire target, whether people, livestock, or crops, at one time (Carus, 2001).

Aerosol Dissemination

Of greatest concern is the possibility that a terrorist might disseminate biological agents as an aerosol cloud. In the context of biological warfare, the aerosol cloud should consist of particles of 1-5 microns (one-millionth of a meter) in size. Particles much larger than 5 microns do not penetrate into the lungs, since they are filtered out by the upper respiratory tract. In addition, they tend to settle out of the air relatively quickly. In contrast, smaller particles do not remain in the lungs, but tend to be breathed out (Sidell, 1997).

Several considerations account for the concern about aerosol delivery. Many diseases are most dangerous when contracted in this fashion. Thus, cutaneous anthrax, which is contracted through the skin, has a case fatality rate of 5 to 20 per cent, though antibiotic treatment is highly effective. In contrast, inhalation anthrax is usually fatal, and if not detected early there is no effective treatment. Similarly, *Y. pestis* is responsible for substantially different diseases, including both bubonic and pneumonic plague. Bubonic plague, generally acquired from the bite of an infected flea, has a case fatality rate of 50 to 60 per cent if untreated, but generally responds to medical treatment. Pneumonic plague is also generally fatal if untreated. Early treatment is essential to save those infected. Pneumonic plague is considered highly contagious, while bubonic plague is not (Carus, 2001; Pile, 1998).

Aerosol transmission also makes it possible to spread biological agents over a large area and thus affect a large number of people in one attack. The Office of Technology Assessment calculated that 100 kilograms of anthrax spread over Washington could kill from one to three million people if disseminated effectively under the right environmental conditions. In contrast, a one-megaton nuclear warhead would kill from 750,000 to 1.9 million people (US Congress, 1993). An alternative set of calculations was prepared for a study by the World Health Organization (WHO). According to estimates prepared by WHO's expert panel, 50 kilograms of dry anthrax used against a city of one million people would kill 36,000 people and incapacitate another 54,000 (WHO, 1970).

Water Contamination

Many pathogens that have had a significant impact on human life, such as *Vibrio cholerae* (which is the organism responsible for cholera) and *Salmonella typhi* (which causes typhoid fever), are water-borne. It is also possible to inject toxic substances into water systems, including toxins. Thus, it is not surprising that some terrorist groups interested in biological agents have targeted municipal water systems (WHO, 1970).

Fortunately, water systems are less vulnerable than often thought. Municipal water systems are designed to eliminate impurities, especially pathogens (Chandler and Landrigan, 2004). As part of this process, communities use filters to remove particles from the water and add chlorine to kill any organisms that remain. In addition, the ID50 for diseases spread through water is often extremely high. One test indicated that only half of persons who ingested 107 (10 million) *S. typhi* organisms became ill (WHO, 1970). According to a Department of Defense biological warfare analyst, it would require "trainloads" of botulinum toxin to contaminate the New York City water supply to cause problems, simply because of the extent to which the toxin is diluted (Erlick, 1990). For all these reasons, infecting a large population through deliberate contamination of water supplies is difficult to accomplish.

Food Contamination

Terrorists also have spread biological agents by contaminating food. In general, only uncooked or improperly stored food is vulnerable, because the heat generated during cooking readily destroys most pathogens and toxins. This implies that a terrorist would need to target foods that are commonly eaten uncooked, or that can be contaminated after being cooked. Alternatively, the terrorists would need to rely on a toxin that survives cooking.

The dangers from deliberate contamination of food have probably grown due to fundamental changes in food distribution systems. The food processing industry has become increasingly centralized, so that contamination introduced at a single facility can affect large numbers of people. In addition, an increasing amount of food is imported, raising the prospect that perpetrators operating in a foreign country might be able to contaminate food eaten in the United States (Chandler and Landrigan, 2004; Rosenberg, 1988).

Human Carrier

The most reliable way to infect someone is to inject the victim with the organisms responsible for causing a disease. Wandering about, purposely infecting others with bacteria or a virus. This method has at least one clear advantage: The agent need not be highly refined because the terrorists need to infect only one individual directly. Moreover, this method is relatively inexpensive and requires no difficult equipment to disseminate the agent. If a highly contagious agent were chosen, and the infected individual able to expose a significant number of people without attracting attention, the attack could lead to widespread illness, and even wider panic. However, the human-as-biological-bomb method is not practical in many ways. This technique avoids most of the technical problems associated with the dissemination of biological agents. Similar techniques can be used with toxins. In addition, some toxins can cause harm even if applied on the skin (Chandler and Landrigan, 2004; Carus, 2001).

Insect Vectors

Many diseases are naturally transmitted by insects. For example, plague is transmitted by certain flea species, yellow fever is carried by one specific mosquito species, *Aedes aegypti*, and typhus is spread by the body louse, *Pediculus humanus corporis*. It is thus not surprising that biological warfare experts have considered insects as potential vectors for biological agents. The Japanese biological warfare program devoted considerable effort to this dissemination route. On at least a few occasions, the Japanese to have used infected fleas to spread plague (Regis, 1999). The United States considered use of mosquitoes to disseminate certain agents, and established a facility to breed the needed mosquitoes (Sidell *et al.*, 1997; Smart, 1989). Insect vectors posed problems: they were difficult to control, and their use was likely to create disease reservoirs in the area where the insects were released.

Consequences of Use

Bioterrorism differs significantly from other forms of terrorism. In the typical terrorist attack, there is immediate evidence that something has happened. In contrast, a biological agent attack is unlikely to generate any visible signatures, and the first evidence of a biological attack is likely to be the onset of disease (Crozier, 1961). Thus, it might be days or even weeks before the consequences of a biological attack become evident. It might be impossible to determine that an outbreak resulted from an intentional act. Numerous efforts have been made to calculate the potential impact that biological weapons might have on population centers. In general, such estimates should be used with considerable caution. Models used to calculate agent effects rely on numerous assumptions regarding the movement and infectiveness of the agent, and the results may be extremely sensitive to changes in those variables. At best, the models provide an illustration of the potential consequences of bioterrorism.

The World Health Organization has prepared the most authoritative estimates of the casualties likely to result from the biological weapon use (WHO, 1970). Although these estimates are highly dependent on the assumptions built into the calculations, they provide a basis for understanding the potential consequences of biological agent use. The impact depends heavily on the nature of the attack, including the method of dissemination, the particular agents involved, the concentration of agent, and (in the case of aerosol dissemination) atmospheric conditions.

Biological Terrorist Agents

The idea of using biological agents as weapons both fascinates and terrifies the public and nothing grabs attention more than the names of agents themselves: anthrax, smallpox, Ebola, etc. Despite their fascination, the public knows little about the specifics of biological agents.

Good coverage of bioterror in general must be informed, and a clear, sophisticated, straightforward discussion of the specific agents is especially critical. To dispel misinformation and clearly delineate the main issues surrounding bioterrorism (Chandler and Landrigan, 2004).

Anthrax

Anthrax Spores

Bacillus anthracis, like many other members of this genus of bacteria, forms tiny spores (endospores) that can travel for considerable distances in the air. Breathing of these spores can lead to pulmonary anthrax (Encyclopedia, The freedictionary.com, Biological warfare, 2010).

The agent (*Bacillus anthracis*-Anthrax) was isolated and characterized by Robert Koch, who described its cultural and morphological characteristics in detail in 1867. It was this organism that fulfilled Koch's postulates for the first time (Chin, 2000). Anthrax is a disease of cattle, goats, and sheep caused by a bacterium, *Bacillus anthracis*. It is rare for humans to be infected. Most infections that do occur are localized to small cuts in the skin whose edges turn black (hence the name "anthracis", after anthracite coal). The disease is deadly for humans because *B. anthracis* produces lethal toxins. Like other members of the *Bacillus* genus, *B. anthracis* produces endospores. An endospore is a tiny dormant cell, a tough package of DNA wrapped in protein that a bacterium makes when times are tough, sort of a "seed" that can persist for centuries, until times improve and the spore germinates to reestablish the anthrax population. Once weaponized, anthrax is easily disseminated, as demonstrated by the attacks of October 2001, and by the Sverdlovsk accident in 1979, which resulted in human fatalities as far as four kilometers away from the release site. However, anthrax is not contagious; only those directly exposed can develop infection (Johnsons, 2010, Logan, 1999).

Although all forms of anthrax are caused by the same bacteria, the effects are very different depending on how the organism enters the body. Inhalational or pulmonary anthrax, which affects the respiratory system, is the most lethal form of exposure to the disease and is therefore currently believed to be the form most likely used in a terrorist attack. Once inhaled, the tiny anthrax spores (one to five microns in size, less than one-twentieth the diameter of a human hair), enter the lungs' alveoli, or air sacs, where blood is oxygenated. Authorities originally believed that at least 10,000 spores were needed to infect a human being, but the October 2001 attacks suggest that much smaller amounts, perhaps just a few thousand-might be enough to cause infection. From the lungs, the infection spreads to the lymph nodes in the chest, and within hours or days, the bacteria begin producing large amounts of a deadly toxin (Chandler and Landrigan, 2004).

Anthrax infection progresses in two phases, the first of which brings flu-like symptoms including fever, nausea, vomiting, aches and fatigue. As with most other biological agents, these symptoms are nonspecific and often resemble the flu so that the initial diagnosis is likely to be incorrect. Health care workers will have to be extremely vigilant to notice a sharp rise in similar cases or in slightly unusual symptoms. The first symptoms usually appear in one to seven days after exposure but in some cases can appear more than a month later. A short recovery-like period sometimes follows the first phase, but the infection progresses to its final phase within two to four days of the onset of symptoms. The second set of symptoms is characterized by respiratory distress and failure, shock and sometimes death. Untreated inhalational anthrax has a fatality rate of approximately 90 percent. Aggressive long-term treatment with antibiotics may reduce the fatality rate to 30 percent. Antibiotic treatment is most successful if begun before the toxin is released, which can occur anywhere from hours to days after exposure. An anthrax vaccine exists, but it is not a treatment option; it is effective only if the first of six inoculations is given at least four weeks before exposure. The vaccine is presently given only to those considered to be at a heightened risk of exposure, including lab workers and certain members of the armed forces. It consists of three injections given two weeks apart, followed by three more injections at six, 12 and 18 months. Annual booster injections are recommended to maintain immunity (Chandler and Landrigan, 2004).

Like most Gram positive bacterial infections, anthrax can be treated effectively with antibiotics if administered early in the infection. Because of worries that an anthrax infection may involve weaponized anthrax that has been made resistant to penicillin, other antibiotics that works differently, ciprofloxacin (CIPRO), iprofloxacin, and doxycycline are the drugs of choice in treating anthrax infections.

An effective vaccine against anthrax was first produced by Louis Pasteur in 1880 using heat-inactivated bacteria. Today's vaccines are a complex broth of proteins filtered from a nonthreatening strain of anthrax. Shots must be repeated for several months to gain full protection. New alternative vaccines based on a genetically engineered version of a single key antigen are going into clinical trials, and are anticipated to produce 95 per cent protection with a single shot (Johnsons, 2010).

Table 4.1: Anthrax through the ages

1500 B.C.	Fifth Egyptian plague, affecting livestock.
1600s	"Black Bane," thought to be anthrax, kills 60,000 cattle in Europe.
1876	Robert Koch confirms bacterial origin of anthrax.
1880	First successful immunization of livestock against anthrax by Louis Pasteur.
1915	German agents acting in the United States believed to have injected horses, mules, and cattle with anthrax on their way to Europe in World War I.
1937	Japan starts biological warfare program in Manchuria, including tests involving anthrax.
1942	England experiments with anthrax at Gruinard Island off the coast of Scotland. The island has only recently been decontaminated.
1943	United States begins developing anthrax bioweapons.
1950s and '60s	U.S. biological weapons program continues after World War II at Fort Detrick, Maryland.
1968	Anthrax bioweapon reported to have been successfully tested at Johnston Atoll in Pacific.
1969	President Richard Nixon ends United StatesÕ biological weapons program.
1972	International convention outlaws development or stockpiling of biological weapons. Russia signs the convention, then secretly undertakes massive expansion of its bioweapons program, making tons of smallpox and anthrax.
1979	Weaponized anthrax aerosol released accidently at a Russian military facility, killing about 68 people.
1990-93	The terrorist group, Aum Shinrikyo, releases anthrax from rooftops in Tokyo, but no one is injured.
1995	Iraq admits it produced 8,500 liters of concentrated anthrax as part of a bioweapons program.
2001	Letters containing milled anthrax are mailed to U.S. news organizations and Congress in the first use of bioweapons by terrorists.

Source: Johnsons, 2010.

Plague

As a member of the family *Enterobacteriaceae*, *Y. pestis* is an oxidase-negative facultative aerobe. Plague is the disease caused by infection with this bacterium (rod shaped). Plague does not receive as much public attention as anthrax or smallpox, but its lethality, contagiousness and infectivity make it one of the most deadly and potentially effective bioweapons. Pneumonic plague (deemed the most likely form of plague to be used in a bioterror attack) has a lethality rate of almost 100 percent if left untreated and approximately 50 percent if treated-high enough to make overcoming the difficulty of acquisition, refinement and dissemination well worth a terrorist's while.

Plague is naturally transmitted to humans either by inhalation or by the bite of a flea that has previously bitten a rodent infected with the bacterium. In the case of a bioterror attack, the bacterium might be released in an aerosolized form into the air. Refining the bacteria to an effective, airborne form that can cause pneumonic plague requires a high degree of technical expertise. Moreover, plague is not extremely stable; it degrades in sunlight or heat but can remain viable for up to a year in the soil. Plague infection in humans can take three forms: pneumonic, bubonic and septicemic. As previously mentioned, pneumonic plague is thought to pose the greatest risk for a bioterror attack because it infects people more easily than the other forms and also is the only form that is contagious. Pneumonic plague results from the inhalation of the bacteria into the lungs or from the spread of infection of the septicemic form. Once inhaled into the lungs, symptoms usually appear after two to four days and include a cough producing bloody mucus, fatigue, fever, diarrhea, nausea and vomiting. The infection can pass from an infected individual to others by coughing. A full pulmonary infection follows the initial symptoms, and death can follow within a day or two if the infection is not treated early and aggressively (Chandler and Landrigan, 2004; Aleksic, 1999).

The bubonic form of plague occurs when an infected flea bites an individual. Instead of infecting the lungs, as in the pneumonic form, bubonic plague infects the lymphatic system. The first symptoms, including weakness, fever and chills, generally appear two to eight days after exposure. These initial symptoms are followed two to four days later with the characteristic and painful swelling of the lymph nodes (called buboes). Untreated, death can follow within a few days. Bubonic plague is not contagious. Septicemic plague can occur when the plague infection enters the bloodstream, leading to internal hemorrhaging and, without prompt treatment, rapid death. Septicemic plague is not contagious (Chandler and Landrigan, 2004). Pneumonic plague can be secondary to bubonic plague. Pneumonic plague presents with fever, malaise, dyspnea, cough, sputum production, and cyanosis (Inglesby, 2000).

Standard and droplet precautions are used with plague. Prophylaxis should be provided for anyone exposed to pneumonic plague. Streptomycin, gentamicin, doxycycline, or ciprofloxacin can be used to treat plague (Inglesby, 2000; Porche, 2002).

Tularemia

Tularemia is a bacterial disease caused by *Francisella tularensis*, an aerobic gram-negative coccobacillus. *Francisella tularensis*is transmitted through tickbites. The natural reservoirs of *F. tularensis* include lagomorphs, rodents, and other animals. Humans can become infected after direct animal contact (often due to hunting, dressing, and consuming infected animals) or via insect vectors such as ticks, biting flies, and mosquitoes (Cross, 2000). Handling or ingestion of undercooked meat from infected animals, drinking contaminated water, or inhalation of dust from contaminated soil, grain or hay.

Direct person-to-person transmission does not occur. The incubation period ranges from 1 to 14 days, but usually is 3 to 5 days (Chin, 2000). Cutaneous tularemia infection presents as a skin lesion with swollen lymph nodes. Ingestion of *Francisella tularensis* produces a throat infection, abdominal pain, diarrhea and vomiting. Inhalation of *Francisella tularensis* produces a fever alone or with pneumonia-like illness. Tularemia infection can be prevented by wearing gloves when skinning or handling animals, thoroughly cooking wild animals, avoiding fly and tick bites, avoiding drinking, bathing, or swimming in untreated water, and avoiding handling sickor dead animals (or using gloves). Gentamycin and tobramycin have reported effectiveness against tularemia infection (Dannis, 2001).

The natural ulceroglandular form of tularemia infection is usually contracted through the bite of an infected tick or fly, or when infected meat comes into direct contact with abraded skin or an open wound. Ulceroglandular tularemia is characterized by the appearance of an ulcer at the infection site and the subsequent swelling of regional lymph nodes. Ulceroglandular tularemia has a lower fatality rate than pneumonic or typhoidal tularemia and is treatable with antibiotics. Tularemia infection can also occur when undercooked, infected meat is consumed (NATO, 2010).

Cholera

Cholera is a diarrheal disease caused by *Vibrio cholera*, a short, curved, gram-negative bacillus. Humans acquire the disease by consuming water or food contaminated with the organism. The organism multiplies in the small intestine and secretes an enterotoxin that causes a secretory diarrhea. When employed as a BW agent, cholera will most likely be used to contaminate water supplies. It is unlikely to be used in aerosol form. Without treatment, death may result from severe dehydration, hypovole mia and shock. Vomiting is often present early in the illness and may complicate oral replacement of fluid losses. There is little or no fever or abdominal pain.

Watery diarrhea can also be caused by enterotoxigenic *E. coli*, rotavirus or other viruses, noncholera vibrios, or food poisoning due to ingestion of preformed toxins such as those of *Clostridium perfringens*, *Bacillus cereus*, or *Staphylococcus aureus*.

There have been six major pandemics of cholera in the last two centuries. During the 1850s London epidemic, John Snow made a map of cholera deaths in London, enabling him to pinpoint a feces-contaminated well on Broad Street as the source of the epidemic (REFERENCE).

Treatment of cholera depends primarily on replacement of fluid and electrolyte losses. This is best accomplished using oral dehydration therapy with the World Health Organization solution (3.5 g NaCl, 2.5 g NaHC03, 1.5 g KCl and 20 g glucose per liter). Intravenous fluid replacement is occasionally needed when vomiting is severe, when the volume of stool output exceeds 7 liters/day, or when severe dehydration with shock has developed. Antibiotics will shorten the duration of diarrhea and thereby reduce fluid losses. Improved oral cholera vaccines are presently being tested. Vaccination with the currently available killed suspension of V. cholera provides about 50 per cent protection that lasts for no more than 6 months. The initial dose is two injections given at least 1 week apart with booster doses every 6 months (WHO, 2010; FDA, 2010; Shapiro, 1999).

Brucella

Isolates of *Brucella* form small, faintly staining gram-negative cocci or short rods. Members of this genus typically behave as slow-growing, fastidious organisms on primary isolation and are nonmotile, non-spore-forming strict aerobes that are catalase positive and usually oxidase and urease positive. DNA-DNA hybridization studies performed in the 1980s suggested that *Brucella* was a monospecific genus, but isolates from human infection are still classified into groups using the species names *Brucella abortus, Brucella melitensis, Brucella suis,* and *Brucella canis,* which reflect the animal species from which strains of the species are likely to be isolated (cattle, goats, pigs, and dogs, respectively.) These groups can be differentiated on the basis of phenotypic traits, such as sensitivity to various dyes, H_2S production, and phage susceptibility.

Humans are infected through contact with animals or animal products that are contaminated with *Brucella. Brucella* is transmitted through eating or drinking contaminated food (unpasteurized milk), inhaling the organism, or entrance of the bacteria through nonintact skin. Direct person-to-person transmission is extremely rare. *Brucella* infection can also be transmitted through breastfeeding. The incubation period ranges from 5 to 60 days (Chin, 2000). Clinical presentation of brucellosis consists of mild flu-like illness to severe central nervous system infection. Chronic symptoms can include recurrent fevers, joint pain, and depression. Prevention consists of pasteurizing milk, cheese, and ice cream. Animal handlers should use gloves. Treatment for brucellosis consists of doxycycline and rifampin (Chin, 2000).

Q Fever

Q fever is a zoonotic disease caused by a rickettsia, *Coxiella burnetii*. The most common animal reservoirs are sheep, cattle and goats. Humans acquire the disease by inhalation of particles contaminated with the organisms. A biological warfare attack would cause disease similar to that occurring naturally.

Following an incubation period of 10-20 days, Q fever generally occurs as a self-limiting febrile illness lasting 2 days to 2 weeks. Pneumonia occurs frequently, usually manifested only by an abnormal chest x-ray. A nonproductive cough and pleuritic c hest pain occur in about one fourth of patients with Q fever pneumonia. Patients usually recover uneventfully.

Q fever usually presents as an undifferentiated febrile illness, or a primary atypical pneumonia, which must be differentiated from pneumonia caused by mycoplasma, legionnaire's disease, psittacosis or Chlamydia pneumonia. More rapidly progressi ve forms of pneumonia may look like bacterial pneumonias including tularemia or plague (WHO, 2010; FDA, 2010; Shapiro, 1999).

Prevention of Q fever consists of standard precautions with gloves for handling items and surfaces contaminated with blood or body fluids.Masks should be worn if there is coughing, and gowns worn to prevent the splash of potentially infected fluids on skin and clothing. Other prevention measures include pasteurizing milk; appropriate disposal of animal placenta, birth products/fluids, aborted animal fetuses; and animal quarantine. Doxycycline is the treatment of choice for Q fever (Porche, 2002).

Vaccination with a single dose of a killed suspension of *C. burnetii* provides complete protection against naturally occurring Q fever and >90 per cent protection against experimental aerosol exposure in human volunteers. Protection lasts for at least 5 years. Administration of this vaccine in immune individuals may cause severe cutaneous reactions including necrosis at the inoculation site. Newer vaccines are under development. Treatment with tetracycline during the incubation period will delay but not prevent the onset of illness (WHO, 2010; FDA, 2010; Shapiro, 1999).

Botulinum

Clostridium botulinum strains and some isolates of clostridia classified as *Clostridium butyricum, Clostridium baratii,* and *Clostridium argentinense* produce a family of seven immunologically distinct potent neurotoxins, designated A through G. Types A, B, E, and F are the usual agents of botulism in humans and cause the clinical syndromes identified as food-borne botulism (an intoxication), wound botulism (infection and toxin production), and botulism caused by toxin production after clostridial colonization of the intestines of infants (infant botulism) or older children and adults. Botulinum toxins spread through the bloodstream and exert their effects at neuromuscular junctions, where they inhibit the release of acetylcholine. The classical presentation of botulism is acute flaccid paralysis that begins in the head and descends symmetrically. When breathing becomes impaired, patients with botulism should be treated with respiratory support. Treatment with antitoxin should also be instituted as soon as the diagnosis is made. Clinicians suspecting botulism should notify the CDC immediately. The CDC provides antitoxin and epidemiological and diagnostic services for botulism cases (allen, 1999; Franz, 1997).

Since botulism toxins are extremely potent, such specimens should be handled with care, employing a biological safety cabinet, disposable gloves, lab coat, and, if needed, face shield. At referral laboratories capable of diagnosing botulism, serum specimens are tested for toxin, as are other specimens, such as vomitus, stool, or tissue debrided from infected wounds. Some specimens may also be cultured for *C. botulinum.* Any isolates are identified and then tested for toxogenicity. While the symptoms, diagnosis, and treatment of food-borne botulism have been well described, less is known about intoxication resulting from inhalation of toxin. Evidence suggests that toxin acquired via an aerosol is less likely to be detected in serum or stool specimens.

Nevertheless, the clinical microbiology laboratory could be called upon to process and refer a variety of specimen types in a suspected bioterrorist event involving aerosolized toxin. However, administration of botulinum antitoxin appears to be extremely effective for treatment of botulism acquired in this manner if antitoxin is given before clinical symptoms become apparent (Chandler and Landrigan, 2004).

Glanders

Glanders (*Burkholderia mallei*), this bacterial disease is highly lethal, killing more than 50 percent of those exposed. Distributed in aerosol form, it produces symptoms within 10 to 14 days, and leads to death from septicemia (blood infection) within seven to 10 days of the onset of symptoms. It can be transmitted from person to person, but at a low rate. It is stable in the environment, and there is no vaccine.

Smallpox

The smallpox (variola) virus is the largest of the animal viruses. The virus particles are brick-shaped to ovoid and measure approximately 300 by 200 by 100 μm. Morphologically, the virus is indistinguishable from the less pathogenic, closely related vaccinia virus, which is one of the best-investigated human viruses. The variola virus contains double-stranded DNA and has a complex structure. Two lipoprotein membrane layers surround the dumbbell-shaped nucleoid. The nucleoid is embedded in an ellipsoid body, forming the thick center of the virion. A double membrane surrounds the virus particle. The variola virus is highly contagious and very virulent, with a case fatality rate of 30 per cent in unvaccinated persons (Handerson, 1999; Klietmann, 2001; Chandler and Landrigan, 2004).

Poxviruses are divided into four different groups. Group 1 comprises variola, vaccinia, cowpox, ectromelia, rabbitpox, and monkeypox viruses. The variola virus exists as one of two strains: variola major causing severe smallpox and variola minor, causing mild smallpox or alastrim. These two strains are immunologically indistinguishable. The first 2 to 4 days of the illness presents as influenza. Skin lesions appear on the face and extremities (not on the palms or soles) progressing from macules to vesicles. The rash appears simultaneously on the body rather than in a progressive nature. The rash forms scabs in 1 to 2 weeks (Klietmann, 2001).

Vaccination against smallpox is performed with the vaccinia virus, which has many antigenic structures in common with the smallpox agent. The vaccinia virus does not exist in nature and is considered a laboratory artifact. It was used worldwide in a live vaccine against smallpox and served as a laboratory model for the poxviruses. The origin of the vaccinia virus remains unclear. It is different from Jenner's cowpox virus and may be a mutant of the variola and alastrim viruses (Handerson, 1999; Klietmann, 2001; Chandler and Landrigan, 2004).

Viral Hemorrhagic Fever

A diverse group of viruses are capable of causing viral hemorrhagic fever syndrome. These include RNA viruses that are members of the *Filoviridae* (Ebola and Marburg viruses), *Arenaviridae* (Lassa fever, Argentine or Junin, Bolivian or Machupo, Venezuelan or Guanarito, and Brazilian or Sabia hemorrhagic fever viruses),

Bunyaviridae (hantavirus, Rift Valley fever, and Congo-Crimean hemorrhagic fever viruses), and *Flaviviridae* (yellow fever and dengue fever viruses). Humans are exposed to these agents by contact with infected animals or via arthropod vectors. The infections caused by this group of viral agents are characterized by vascular damage and altered vascular permeability. Symptoms commonly include fever and myalgias, prostration, hemorrhages in mucous membranes, and shock. Viral hemorrhagic fevers cause high morbidity, and in many cases high mortality rates are observed. Treatment consists mostly of supportive measures, although the antiviral agent ribavirin seems to be useful for treatment of infection with certain agents such as Lassa fever virus, Junin, Bolivian, and Congo-Crimean hemorrhagic fever viruses, and Rift Valley fever virus (Franz, 1997).

Enterotoxin B

Enterotoxin B (produced by *Staphylococcus*) This toxin, produced by staph bacteria, can cause illness within a few hours to six days, but has a very low lethality rate (less than 1 percent), and illness only lasts a few hours. It is stable in the environment, and can even survive freezing. There is no vaccine (Chandler and Landrigan, 2004).

Epsilon Toxin

Epsilon toxin (produced by *Clostridium perfringens*) A common cause of food poisoning, especially from improperly cooked beef or chicken (Chandler and Landrigan, 2004).

Ricin

Ricin (toxin from *Ricinus communis,* or castor beans) This naturally produced toxin could be spread in aerosol form, and can produce symptoms within hours to days. It is stable in the environment, and can be fatal within 10 to 12 days (Chandler and Landrigan, 2004; Ivanhoe Newswire, 2010].

Viral Encephalitis

Viral encephalitis (alphaviruses, such as Venezuelan equine encephalitis, Eastern equine encephalitis, Western equine encephalitis) A viral disease, it is unstable once dispersed, so only those directly contaminated in the initial attack will become ill. It has a low lethality rate, producing symptoms in one to six days (Chandler and Landrigan, 2004).

Hantavirus

Hantavirus carried by rodents and mostly transmitted through their droppings, this virus was responsible for an outbreak of disease in Arizona and New Mexico in 1993. It causes Hantavirus Pulmonary Syndrome (HPS), which has now been identified in eight other countries besides the United States (all in the Americas). It has been fatal in 45 percent of reported cases, causing death through pulmonary edema and respiratory distress. There is evidence of human-to-human transmission (Chandler and Landrigan, 2004).

Nipah Virus

Nipah Virus, a "new" virus, it was discovered in Malaysia in 1999, closely related to Hendra virus, discovered in Australia. Both of these are *Paramyxoviridae*. It infects both animals (mostly pigs) and humans, and has a high mortality rate (50 percent). It begins with flu-like symptoms, and can progress to encephalitis (brain inflammation) and coma (Chandler and Landrigan, 2004).

Other Possible Biological Weapons

A variety of other agents have been developed, and in some cases even stockpiled, by some nations as possible biological weapons. Others have been named by the World Health Organization and other agencies as possible bioweapons. Some of these are specific examples that fall within categories already included in the CDC list. Since many of these agents were actually developed for warfare by at least one nation, their potential as weapons is not just theoretical. In addition to those described previously, these include the following.

Aflatoxin

Although not usually considered a candidate for a biological weapon, primarily because its effects tend to be very long-term (cancer and respiratory disease that take years to develop), aflatoxin was in fact produced on a large scale as a weapon, mounted in munitions and missiles, and perhaps used against Iranian civilians, by Iraq in the 1980s and 1990s. Although technically a chemical weapon, it is produced by fungi and therefore sometimes classified with biological weapons.

Multi-Drug-Resistant Tuberculosis

Tuberculosis, now rare in most countries, remains a major killer in developing nations. It kills about 2 million people per year. Normally, antibiotic treatment cures 95 percent of cases, but the resistant strains may require aggressive chemotherapy treatment for up to two years. Recent studies show that more than 4 percent of new cases are of the multi-drug resistant type (MDR-TB).

Tricothecene Mycotoxins

This natural toxin could be distributed in an aerosol form, and begins to produce symptoms within two to four hours (one of the fastest-acting biological agents) which can persist for days to months. It is moderately lethal, and extremely stable-at room temperature, it can remain dangerous for years. There is no vaccine.

Bacteria

Trench fever (*Bartonella quintana*) and scrub typhus (*Orientia tsutsugamushi*).

Fungi

Coccidioidomycosis (*Coccidiodes immitis*) and histoplasmosis (*Histoplasma capsalatum*).

Protozoa

Naegleriasis (*Naeglaeria fowlerii*), toxoplasmosis (*Toxoplasma gondii*), and schistosomiasis (*Schistosoma*).

Viruses

Hantaan/Korean hemorrhagic fever, sin nombre, Crimeo-Congo hemorrhagic fever, lymphocytic choriomeningitis, Junin (Argentine hemorrhagic fever), Machupu (Bolivian hemorrhagic fever), tick-borne encephalitis, Russian spring-summer encephalitis, Omsk hemorrhagic fever, Japanese encephalitis, Chikungunya, O'nyong-nyong,monkeypox, white pox (a variant of *variola*), and influenza (Chandler and Landrigan, 2004).

Agricultural Bioterrorism

Although most concerns have focused on human pathogens, a biological attack against crops or livestock is also considered a possibility. A study by the National Research Council concluded that such an attack would almost certainly not be capable of producing famine or widespread malnutrition in the United States, but it could nevertheless have a severe economic impact. One study showed that an outbreak of foot and mouth disease (a cattle disease) in California alone, even if contained within three months, could cause a loss of $6 billion to $13 billion. Some of the plant and animal diseases considered to pose the greatest threat (Jaiswal and Sundar, 2010):

Avian Flu

Avian flu-Can have a serious economic impact. An outbreak in the United States in the mid-1980s resulted in the destruction of 17 million birds at a cost of $65 million. Found in humans only in Vietnam and Thailand as of early 2004.

Foot and Mouth Disease

Foot and mouth disease (FMD)-Can have serious economic impact, though its effects on cattle are relatively minor. To avert potential losses from exports, Taiwan spent $4 billion in an attempt to eradicate the disease in 1997, without success.

Karnal Bunt of Wheat

Karnal bunt of wheat-Even though it has little impact on wheat productivity, 80 countries ban wheat imports from areas where the disease has been found. A single outbreak in Arizona in 1996, probably because of an accidental introduction across the border from Mexico, posed a threat to that state's $6 billion in wheat exports, and caused over $100 million in actual losses, in addition to $60 million spent on control efforts.

Mad Cow Disease

Mad Cow Disease (MCD)-Though its actual impact on cattle is relatively minor, outbreaks have had devastating economic impact, as in England and much of Europe in 1996. In Japan in 2001, the detection of just three cases of the disease led to a 50 percent reduction in beef sales. So far, there have only been two documented U.S. cases, in late 2003, which were both traced to cows from a Canadian herd.

Swine Fever

An outbreak in the Netherlands in 1997-'98 was caused by swine imported from Germany, and illustrated how a rapid outbreak could follow a single introduction into a previously uninfected country. Other plant pathogens that have been named as possible weapons include stem rust (in wheat), sorghum ergot and barley stripe rust

Sorghum ergot, and barley stripe rust. There is no accepted list of the most likely biological agents that might be used against crops, and some researchers have suggested that developing such a list is an important first step toward monitoring to detect a possible attack. At present, an outbreak might progress to a point at which control will be difficult and expensive before being detected. Some other nations, with a less diverse agricultural base than the United States, could suffer more serious health and economic effects from such an attack (Jaiswal and Sundar, 2010).

Trendy Bioweapons

Recent scares about anthrax in the United States have brought forth the topic of bioterrorism. Such biological weapons are often classified along with chemical and nuclear weapons as Weapons of Mass Destruction (WMD). However, (Biological Weapons) BW are dissimilar from the other two in the fact that they are much cheaper to produce, and their effects are maximally felt only after a certain gestation period.

The evolution of chemical and biological weapons can broadly be categorized into four phases. In the first phase, chemicals like chlorine and phosgene were used in Ypres in World War I. The second phase ushered in the use of nerve agents *e.g.* tabun, a cholinesterase inhibitor, and the alleged use of biological weapons in World War II. The third stage was constituted by use of lethal chemical agents like Agent Orange. This phase also included the use of the new group of Novichok and mid spectrum agents that posses the characteristics of chemical and physiologically active compounds. The fourth phase coincides with the era of biotechnology and the extensive use of genetic engineering.

The new tools of biotechnology and genetic engineering have enabled the creation of pathogens deadlier than ever before. The Soviets were reputed to have created a new strain of anthrax which was immune to most antibiotics, including the much vaunted Ciproflavoxin. These methods have also enabled the creation of so called "Ethnic Bombs". In the fall of 1998, there was a report that the white South African government had developed a program that would specifically kill blacks. A rumor too surfaced (in the English Press) that Israel was working on a BW that would specifically harm Arabs carrying certain genes.

Now, we would like to discuss a protein toxin called ricin, which is a protein toxin that is readily produced from castor beans (*Ricinus communis*), which are ubiquitous throughout the world. It kills by destroying the ribosomes which synthesize proteins in the human body. Acting as a slow poison, it eventually causes a total collapse as body proteins are not replaced. The structure and mechanism of action of ricin is well known, so it is a prime candidate for genetic manipulation. Ricin is already being investigated for its "magic bullet" properties as an agent that might selectively destroy cancer cells. This same technology could easily be applied to improving its BW- capacity. For example, if ricin is chemically bound to antibodies that only bind to a certain type of cancer cell, the attached ricin should only kill the targeted cancer cells and no other cells. The same principle could be used to specifically target an enemy; in theory one could be specific enough to use this process to target a single individual for assassination.

The threat of terrorists using biological weapons is no longer a remote one. The new tools of technology have enabled them to create and modify strains to be more virulent than ever before. Thus it is imperative that we prepare well in advance to counter this new age threat (Ainscough, 1989; Marx, 1989; USGAO, 1999; Johnsons, 2010).

Conclusion

At this point in the discussion of the complex issue of bioterrorism, we are in a position to raise more questions than answers. However, our intent is to contribute to a fruitful dialogue which will stimulate the process to generate answers that will be the building blocks for our preparedness and the nation's defense against the threat of bioterrorism. The threat of bioterrorism is real and looming before us. The involvement of clinical microbiologists and other health care professionals in preparing for a bioterrorist act should extend beyond the institution of protocols and plans to be followed in the wake of such an event. To be prepared in a responsive and responsible way, we need a paradigm of scenario planning in which we look at a range of possibilities and countermeasures rather than construct a linear and reactive strategic plan. This demands that, in our collective imagination, we move the previously unthinkable into the realm of possibility in order to develop a realistic response strategy. We will have to integrate many conflicting issues and satisfy conflicting needs through compromises that seek to find the second-best solutions. This will still be smarter than having no solutions at all. The answer is to reconsider the present rather than to prophesy the future based on vague assumptions. The time has come to get prepared and develop an integrated policy. Recently Chairman of the Joint Chiefs of Staff instructions revealed joint biological weapon defense capabilities (CJCSI, 2010) proving we are getting ready against new age war.

References

Ainscough, M. J. (1989). "The Gathering Biological Warfare Storm". in *Next generation bioweapons,* available at http://www.bibliotecapleyades.net/ciencia/ciencia_virus08.htm (accessed 22 April, 2010)

Aleksic, S., and J. Bockemuhl (1999). "Yersinia and other Enterobacteriaceae," in P. R. Murray, E. J. Baron, M. A. Pfaller, F. C. Tenover, and R. H. Yolken (Eds.), *Manual of clinical microbiology, 7th ed.* American Society for Microbiology, Washington. D.C. pp. 483-496.

Allen, S. D., C. L. Emery, and J. A. Siders. (1999). "*Clostridium,*" In P. R. Murray, E. J. Baron, M. A. Pfaller, F. C. Tenover, and R. H. Yolken (Eds.), *Manual of clinical microbiology, 7th ed.* American Society for Microbiology, Washington, D.C. pp. 654-671.

Benenson, A. S. (1995). *Control of Communicable Disease Manual, 16th edition* (Washington, DC: American Public Health Association).

Carus, W. S. (2001) *WORKINGPAPER: Bioterrorism and Biocrimes; The Illicit Use of Biological Agents Since 1900*. Center for Counterproliferation Research, National Defense University, Washington, D.C.

CDC.3, (1998). *Preventing emerging infectious diseases: a strategy for the 21st century*. Atlanta, Georgia: U.S. Department of Health and Human Services.

Chandler, D., and I, Landrigan (2004). *Bioterrorism: A Journalist's Guide to Covering Bioterrorism, second edition*, Carnegie Corporation, New York, USA.

Chin, J. E. (2000). *"Control of Communicable Diseases Manual, 17ᵗʰ Edition"* American Public Health Association, Washington, D.C.

Christopher, G. W., T. J. Cieslak, J. A. Pavlin, and E. M. Eitzen, Jr. (1997). "Biological warfare: a historical perspective", *Journal of the American Medical Association,* Vol. 278, pp. 412-417.

CJCSI, (2010). "Joint Biological warfare Defense Capabilities," in *Chairman of the Joint Chiefs of the Staff Instruction,* available at http://www.dtic.mil/cjcs_directives/cdata/unlimit/3112_01.pdf (accessed 22 April, 2010).

Cross, J. T. Jr., and R. L. Penn. (2000). *"Francisella tularensis* (tularemia)", in Mandell, G. L., J. E. Bennett, and R. Dolin (Eds.), *Principles and practice of infectious diseases, 5th ed.* Churchill Livingstone, Philadelphia, pp. 2393-2402.

Crozier, D., W. D. Tiggertt, and J. W. Cooch, (1961). "The Physician's Role in the Defense against Biological Weapons," *Journal of the American Medical Association,* Vol. 175, pp. 4-8.

Debashis, S. (2001). *Biological Weapons and Bio Terrorism: Bio terrorism is real and is here,* Paper no. 342, available at http://www.southasiaanalysis.org/per cent 5Cpapers4 per cent 5Cpaper342.html (accessed 22 April, 2010).

DOD (2009). Joint Publication 1-02, DOD Dictionary of Military and Associated Terms, 546, available at http://www.dtic.mil/doctrine/dod_dictionary/data/t/7591.html (accessed 22 April, 2010).

DOD (United States Department of Defense) (1992). *Office of the Secretary of Defense, Conduct of the Persian Gulf War,* Final Report to Congress (Washington, D.C.: US Government Printing Office).

Eickhoff, T.C. (1996). Airborne disease, including chemical and biological warfare, *American Journal of Epidemiology,* Vol. 144, pp. S39-S46.

Eitzen E. M. (1997). "Use of biological weapons," in Sidell, F. R., Ernst, T. and D. R. Franz (Eds.) *Medical Aspects of Chemical and Biological Warfare,* (Washington DC: Office of the Surgeon General at TMM Publications), 446.

Encyclopedia, The freedictionary.com, 2010. "Biological warfare", available at http://encyclopedia2.thefreedictionary.com/biological+warfare (accessed 22 April, 2010)

Encyclopedia, The freedictionary.com, 2010. "Chemical warfare", available at http://www.thefreedictionary.com/chemical+warfare (accessed 22 April, 2010).

Encyclopedia, The freedictionary.com, 2010. "Nuclear weapon", available at http://encyclopedia.thefreedictionary.com/nuclear+weapon (accessed 22 April, 2010)

Encyclopedia, The freedictionary.com, 2010. "Radiological weapon", available at http://encyclopedia.thefreedictionary.com/radiological+weapon (accessed 22 April, 2010).

Encyclopedia, The freedictionary.com, 2010. "Weapon of mass destruction", available at http://www.thefreedictionary.com/weapon+of+mass+destruction (accessed 22 April, 2010).

Erlick, B. (1990). "Global Spread of Chemical and Biological Weapons", in *United States Senate, Committee on Governmental Affairs, Permanent Subcommittee on Investigations,* Washington, D.C.: Government Printing Office, p. 38.

FBI (2010). The National Foreign Intelligence Program, Economic Espionage, available at http://baltimore.fbi.gov/nfip.htm (accessed 22 April, 2010).

FDA, (2010) The"Bad Bug Book" Foodborne Pathogenic Microorganisms and Natural Toxins Handbook, U.S. Food and Drug Administration, Center for Food Safety and Applied Nutrition, available at http://www.fda.gov/Food/FoodSafety/FoodborneIllness/FoodborneIllnessFoodbornePathogensNaturalToxins/BadBugBook/default.htm (accessed 22 April, 2010).

Franz, D. R., P. B. Jahrling, A. M. Friedlander, D. J. McClain, D. L. Hoover, W. R. Bryne, J. A. Pavlin, G. W. Christopher, and E. M. Eitzen. (1997). "Clinical recognition and management of patients exposed to biological warfare agents", *Journal of the American Medical Association,* Vol. 278, pp. 399-411.

Harris, S. H. (1992). "Japanese biological warfare research on humans: a case study on microbiology and ethics", *Annals of the New York Academy of Sciences.* Vol. 666, pp. 21-52.

Harris, S. H. (2002). "Japanese biological warfare 1932-45, and American coverup", in Factories of Death, Taylor and Francis, eLibrary.

Henderson, D. A., Inglesby, T. V., Bartlett, J. G., Ascher, M. S., Eitzen, E., Jahrling, P. B., Hauer, J., Layton, M., McDade, J., Osterholm, M. T., O'Toole, T., Parker, G., Perl, T., Russell, P. K., Tonat, K. (1999). "Smallpox as a biological weapon: Medical and public health management", *Journal of the American Medical Association,* Vol. 281 No. 22, pp. 2127-2137.

Dennis, D. T., T. V. Inglesby, D. A. Henderson, J. G. Bartlett, M. S. Ascher, E. Eitzen, A. D. Fine, A. M. Friedlander, J. Hauer, M. Layton, S. R. Lillibridge, J. E. McDade, M. T. Osterholm, T. O'Toole, G. Parker, T. M. Perl, P. K. Russell, K. Tonat, and Working Group on Civilian Biodefense, (2001). "Tularemia as a biological weapon: Medical and public health management," *Journal of the American Medical Association,* Vol. 285 No. 21, pp. 2763-2773.

Inglesby, T. V., D. T. Dennis, D. A. Henderson, J. G. Bartlett, M. S. Ascher, E. Eitzen, A. D. Fine, A. M. Friedlander, J. Hauer, M. Layton, J. E. McDade, M. T. Osterholm, T. O'Toole, G. Parker, T. M. Perl, P. K. Russell, M. Schoch-Spana, K., and Tonat, (2000). "Plague as a biological weapon: Medical and public health management", *Journal of the American Medical Association,* Vol. 283 No. 17, pp. 2281-2290.

Ivanhoe Newswire, (2010). "Ricin: The Next Bioterrorism Weapon?", available at http://www.ivanhoe.com/channels/p_channelstory.cfm?storyid=24025 (accessed 22 April, 2010).

Jaiswal, P., and S. Sundar, (2010) "Designer weapons-How far away are they?," available at http://www2.dupont.com/Our_Company/en_IN/assets/downloads/2002_Senior_3.pdf (accessed 22 April 2010).

Johnson, G. (2010). "The Battle against Infectious Disease," in *To defeat an infectious disease, you must control its vector,* available at http://www.txtwriter.com/Backgrounders/Bioterrorism/bioterror2.html (accessed 22 April, 2010).

Johnsons, G. (2010) *"A Closer Look at Anthrax",* available at http://www.txtwriter.com/Backgrounders/Bioterrorism/bioterror5.html (accessed 22 April, 2010).

Klietmann, W. F., and K. L. Ruoff, (2001). "Bioterrorism: implications for the clinical microbiologist", *Clinical Microbiology Reviews.* Vol. 14 No. 2, pp. 364-381.

Logan, N. A., and P. C. B. Turnbull (1999) "Bacillus and recently derived genera," in P. R. Murray, E. J. Baron, M. A. Pfaller, F. C. Tenover, and R. H. Yolken (Eds.), *Manual of clinical microbiology, 7th ed.* American Society for Microbiology, Washington. D.C. pp. 357-369.

Marx, J. L. (1989). *"A Revolution in Biotechnology",* Cambridge University Press, Cambridge.

Mayer, T.N. (1995). The Biological Weapon: A Poor Nation's Weapon of Mass Destruction, Chapter 8, available at http://www.airpower.maxwell.af.mil/airchronicles/battle/chp8.html (accessed 22 April, 2010).

Meselson, M. (2001). "Note Regarding Source Strength," *The ASA newsletter,* available at http://www.asanltr.com/newsletter/01-6/articles/016a.htm (accessed 22 April, 2010)

Meselson, M., J. Guillemin, M. Hugh-Jones, A. Langmuir, I. Popova, A. Shelokov and O. Yampolskaya, (1995). "The Sverdlovsk Anthrax Outbreak of 1979," *Science,* Vol. 266 No. 5188, pp. 1202-1208.

NATO (2010) handbook on the medical aspects of NBS defensive operations Part-II – Biological Guidelines for the Surveillance and Control of Anthrax in Humans and Animals, World Health Organization WHO/EMC/ZDI/98.6, available at http://www.fas.org/programs/ssp/bio/resource/agents.html (accessed 22 April 2010).

Pile, J. C., J. D. Malone, E. M. Eitzen, and A. M. Friedlander, (1998). "Anthrax as a Potential Biological Warfare Agent," *Archives of Internal Medicine,* Vol. 158, pp. 429-434.

Porche, D. J. (2002). "Biological and Chemical Bioterrorism Agents", *Journal of the association of nurses in AIDS care*, Vol. 13 No. 5, pp. 57-64.

Poupard, J. A., and L. A. Miller (1992). "History of biological warfare: catapults to capsomeres", *Annals of New York Academy of Sciences*, Vol. 666, pp. 9-20.

Regis, E. (1999). *"Biology of Doom: The History of America's Secret Germ Warfare Project"*, New York: Henry Holt and Company, pp. 18-19, 112-113.

Rosenberg, J. (1988). "Attack of the killer shellfish, raspberries, hamburgers, milk, poultry, eggs, vegetables, potato salad, pork", *American Medical News*, pp. 12-15.

Shapiro, D. S., and J. D. Wong. (1999). "Brucella", In P. R. Murray, E. J. Baron, M. A. Pfaller, F. C. Tenover, and R. H. Yolken (Eds.), *Manual of clinical microbiology, 7th ed.* American Society for Microbiology, Washington. D.C. pp. 625-631.

Sharma, R. (2001). "India wakes up to threat of Bioterrorism", *British Medical Journal news* Vol. 323, pp. 714, available at http://www.bmj.com/cgi/content/extract/323/7315/714/a (accessed 22 April, 2010).

Sidell, F. R. and D. R. Franz, (1997) "Overview: Defense Against the Effects of Chemical and Biological Warfare Agents," in Sidell, F. R., Ernst, T. and D. R. Franz (Eds.) *Medical Aspects of Chemical and Biological Warfare*, (Washington DC: Office of the Surgeon General at TMM Publications), 437-450.

Sidell, F. R., Ernst, T. and D. R. Franz (1997) *Medical Aspects of Chemical and Biological Warfare*, (Washington DC: Office of the Surgeon General at TMM Publications),

Smart, J. K. (1989) "History of Chemical and Biological Warfare: An American Perspective," in Textbook of Military Medicine: Medical Aspects of Chemical and Biological Warfare (Washington, DC: Office of the Surgeon General, US Department of the Army, 1989), 9-86.

Tucker, J. B. (1993). "Lessons of Iraq's Biological Warfare Programme", *Contemporary Security Policy*, Volume 14 Issue 3, pp. 229-271.

US Army, (1997). *"U.S. Army Activity in the U.S. Biological Warfare Programs"*, vol. 2 (Washington, D.C.: US Government Printing Office).

US Congress, (1993). *"Office of Technology Assessment, Proliferation of Weapons of Mass Destruction: Assessing the Risks"*, OTAISC- 559, Washington, DC: Government Printing Office, pp. 53-54.

USGAO, (1999). "Combating Terrorism: Need for Comprehensive Threat and Risk Assessments of Chemical and Biological Attacks," *The United States General Accounting Office (GAO),*available at http://www.gao.gov/archive/1999/ns99163.pdf (accessed 22 April 2010).

Watson, A. and D. Keir, (1994) "Information on which to base assessments of risk from environments contaminated with anthrax spores," *Epidemiology and Infection*, Vol. 113, pp. 479-490.

WHO (2010) recommended guidelines for epidemic preparedness and response: Ebola Haemorrhagic Fever (EHF) WHO/EMC/DIS/97.7, available at http://www.who.int/csr/resources/publications/ebola/WHO_EMC_DIS_97_7_EN/en/(accessed 22 April, 2010).

WHO, (1970). *"Health Aspects of Chemical and Biological Weapons"*, Geneva:World Health Organization, Geneva.

Zkea Archives, (2010). Biological Warfare; Emerging Disease, Animals, in *Category: Emerging Diseases: Biological Terrorism,* available at http://www.zkea.com/archives/archive04003.html (accessed 22 April, 2010).

Case Studies

5

Space Application in Disaster Assessment and Mitigation: Case of Haryana State

B.S. Chaudhary

Introduction

Disasters are becoming more and more frequently occurring events in India. Haryana is also suffering from a number of disasters such as Floods, desertification, soil erosion, land degradation etc. Earthquakes are also frequently occurring but of small magnitude so are not causing much concern and damage. Most of the damage in Haryana is due to floods. The present paper deals with the Remote Sensing applications in mapping and management of floods which occurred in September 1995 in the state of Haryana. Haryana witnessed floods in the years 1978, 1988, 1993 and 1995. The floods of September 1995 were most severe as major portion of Haryana was affected except few northern districts of Panchkula, Ambala and Yamuna Nagar. Satellite data from IRS, LANDSAT and SPOT were used for carrying out the study. Visual image interpretation was carried out for mapping flood affected areas. The flood affected areas were divided in to standing water and receded water categories. It was found that heavy downpour along with poor inland drainage, presence of localized depressions, breaching of canals and diminished carrying capacity of the water channels added to the woes. Back flush in the Drain No. 8 also caused flooding in the some areas of Gurgaon and Jhajjar districts. The paper also suggests suitable recommendation measures for mitigating the ill effects of floods in the state.

Floods have become a frequently occurring annual event in many parts of the major river basins almost all over the globe causing great havoc and extensive losses by damaging, agricultural crops, property, human and animal life and thereby causing environmental imbalances also at places. The magnitude of loss all over the globe runs into over 2 billion dollars every year.

India is also not exception to that. India is a vast country with wide variations in climate and topography. Some major rivers such as Ganga, Brahmputra, Indus and their tributaries, Narmada, Godavari and Mahanadi etc traverse it. Many of these major rivers cause floods during the southwest monsoon period but during lean period they dwindle to mere trickles. Due to widespread and heavy rainfall during the monsoon months, floods of varying magnitude occur all over the country during June to September. The magnitude depends upon the intensity of the rainfall, its duration and also the ground conditions at the time of rainfall. As per the National Flood Commission (1990), about 40 million ha of land, which is about one eighth of the geographical area of the country, is prone top recurring floods.

Because of the heavy floods and discharges, various flood prone rivers and their tributaries bring down exceptionally large volumes of sediments which causes the aggravation of river beds and bring changes in the river courses. Quite often, the landslides and earth tremors/earthquakes aggravate the flood situation. The agricultural crops and utilities in the flood plains suffer heavy damages when vast areas are inundated due to one or more causes like water spilling over the river banks, synchronization of high stages of floods in the main river and its tributaries, drainage congestion at outfall of tributaries, silts deposition etc.

Every year about 5 to 6 million ha of crop area is affected due to floods and the total value of damage to crops, houses and utilities is estimated to Rs. 2,500 crores (National Flood Commission, 1990). After the disastrous floods of the year 1954, a national programme for controlling the floods was conceived. During the last 35 years, efforts have been made by the central and state governments to reduce the flood damages by constructing the embankments, reservoirs etc. These efforts have provided protection to about 30 per cent of the flood affected/flood prone areas. The flood-affected area in 1987 was about 11.4 million ha compared to 6 million ha in 1981 (Rao, 1989). As the floods can not be totally controlled and its is not possible to provide protection against all the magnitudes of floods, implementation of appropriate flood management measures both the structural and non-structural have become absolutely necessary. Haryana has also suffered from Floods in 1978, 1988, 1993 and 1995. The floods of September 1995 were most severe as almost two third of Haryana was affected except few northern districts. The present study deals with the mapping of 13 flood-affected districts of Haryana state. The causes of floods vary from Heavy downpour, poor inland drainage, drainage congestion, water spill resulting to canal breach, reduction in effective water way of rivers/other water channels to back flush in drain No. 8. As Ghaggar is the main river traversing Haryana and parts of Punjab and is also a cause of worry due to floods, due to reduction in effective water way of Ghaggar.

Remote Sensing and Flood Mapping

In order to scientifically plan and properly execute the appropriate flood control and anti erosion activities, it is very essential to acquire timely and reliable information about the flooded areas, river behaviors and configuration prior to floods, during the floods and after floods. Such type of information is very difficult to acquire through conventional ground surveys which are very arduous, vague, time consuming and expensive. The advent of satellite remote sensing technology has helped in solving the problems of mapping, monitoring and management of the flooded and flood prone areas.

The data acquired from the space borne remote sensing satellite have been used in India for more than three decades. It has been found that the space borne satellite imageries and digital data can provide a valuable information for carrying the flood management measures for moderating floods, modifying the susceptibility of flood plains for flooding and for reducing the impact of floods.

Justification of Remote Sensing and GIS

The satellite technology by virtue of their remote sensing and data transmitting capabilities provides a comprehensive multi-date and multi-spectral information on the aerial extent of flooded areas, mapping, monitoring and management of these areas, despite the limitations of cloud cover, especially during flood time. Conventional technology can't be used to prepare such maps on near real time basis due to obvious inherent problems. Remote Sensing Technology, which is not only fast and accurate but also cost effective; is a powerful tool for survey, mapping and regular monitoring in this type of situations. The remote sensing technology can effectively used in the flood studies especially relating to:

☆ Flood inundated and drainage congested areas

☆ Extent of damage to crops, structures *etc.*

☆ River configuration and silt deposits

☆ Vulnerable areas for erosion

Advent of satellite remote sensing and geographic information systems has opened new vistas in the field of flood mapping and management. This is due to the fact that earth-observing devices provide most up to date, accurate, unbiased and detailed spectral, spatial and temporal information on the conditions of natural resources.

GIS is defined as an information system that is used to input, store, retrieve, manipulate, analyze and output geographically referenced data or geospatial data in order to support decision making for planning and management of land use, natural resources, environment, transportation, urban facilities, and other administrative records (Murai, 1996). Use of GIS not only helps in manipulating the data and overlaying in a suitable manner so that different composite maps can be prepared by integrating data available in a variety of thematic information layers.

Review of Literature

Satellite applications in mapping and management of floods have been successfully demonstrated in India and Abroad. One of the basic problems faced in the application of optical remote sensing data is the presence of cloud cover which is to extent removed by the more frequent revisit by a combination (Indian and foreign) of satellite for the same area. Another development is the accessibility/availability of Microwave data, which can be used as a supplementary or complimentary to the existing optical data. Some authors have also used the data fusion techniques by applying various algorithms for taking advantage of Spectral and Spatial resolutions of both optical and microwave data. Some work has also been carried in the direction of rainfall-runoff estimates in different catchments for improving the flood forecast warning systems. The availability of high resolution satellite data like PAN, SRTM and ASTER has not only improved the DEM generation with more accuracy but has also helped in understanding these phenomenon in a better way. The work on Satellite applications in conjunction with GIS and GPS is an active area of research in India and Abroad (Florenzano, 1989; De Brouwer, 1994; Rao, 1989; Ramamoorthy, 1989; Sharma 1989; Nageshwar Rao and Jayraman, 1992; chaudhary et al 1999; Rao, 2000; Bhanumurthy et al 2004). Navalgund, 2006 has discussed the potentialities of Geoinformatics in the field of disaster management citing RS, GIS, GPS and Photogrammetry applications in Floods, Cyclone warning, desertification, landslides, earthquakes and Tsunamis monitoring. It is understood that the RS data in conjunction with GIS will help in even calculating the loss to the property, agricultural fields, infrastructure facilities and even per capita losses.

Study Area

Haryana state covers a total area of 44,212 sq kms and likes between 27°, 35' to 31°, 55.5' N latitude and longitudes 74°, 22.8' and 77° 35.6' E longitudes. The state is bounded by natural features as Siwaliks in North, Aravallis in South, Yamuna River in East and Ghaggar in West. The area is covered by Survey of India toposheets Nos. 44K, 44O, 44P, 53C, 53D, 53F, 53G, 53H, 54E. Barring the Hilly portions, most of state lies between 200 to 300 meters above mean sea level. The average slope of the terrain is northeast to southwest however the slope in northern portion is towards south and in south it is towards north making it almost saucer shaped physiography. It mainly occupies Indo-Gangetic water divide and majority is covered by Indo-Gangetic alluvium. Ghaggar, Tangri, Markanda, Saraswati and Chautang originate in Siwaliks in the north, Yamuna is the only perennial river making its eastern boundary, Sahibi, Krishnawati and Dohan non-perennial rivers originate in the south in Aravalli hills of Rajasthan. Haryana is an agricultural state with 88 per cent of the total geographical area under cultivation and cropping intensity of more than 150 per cent. Dominant food crops in the state are wheat, paddy, maize and gram. The thirteen districts of Haryana covered under the flood mapping exercise are Bhiwani, Faridabad, Gurgaon, Hisar, Jind, Kaithal, Karnal, Kurukshetra, Panipat, Rewari, Rohtak, Sirsa and Sonipat. The satellite coverage map of the state is shown in Figure 5.1.

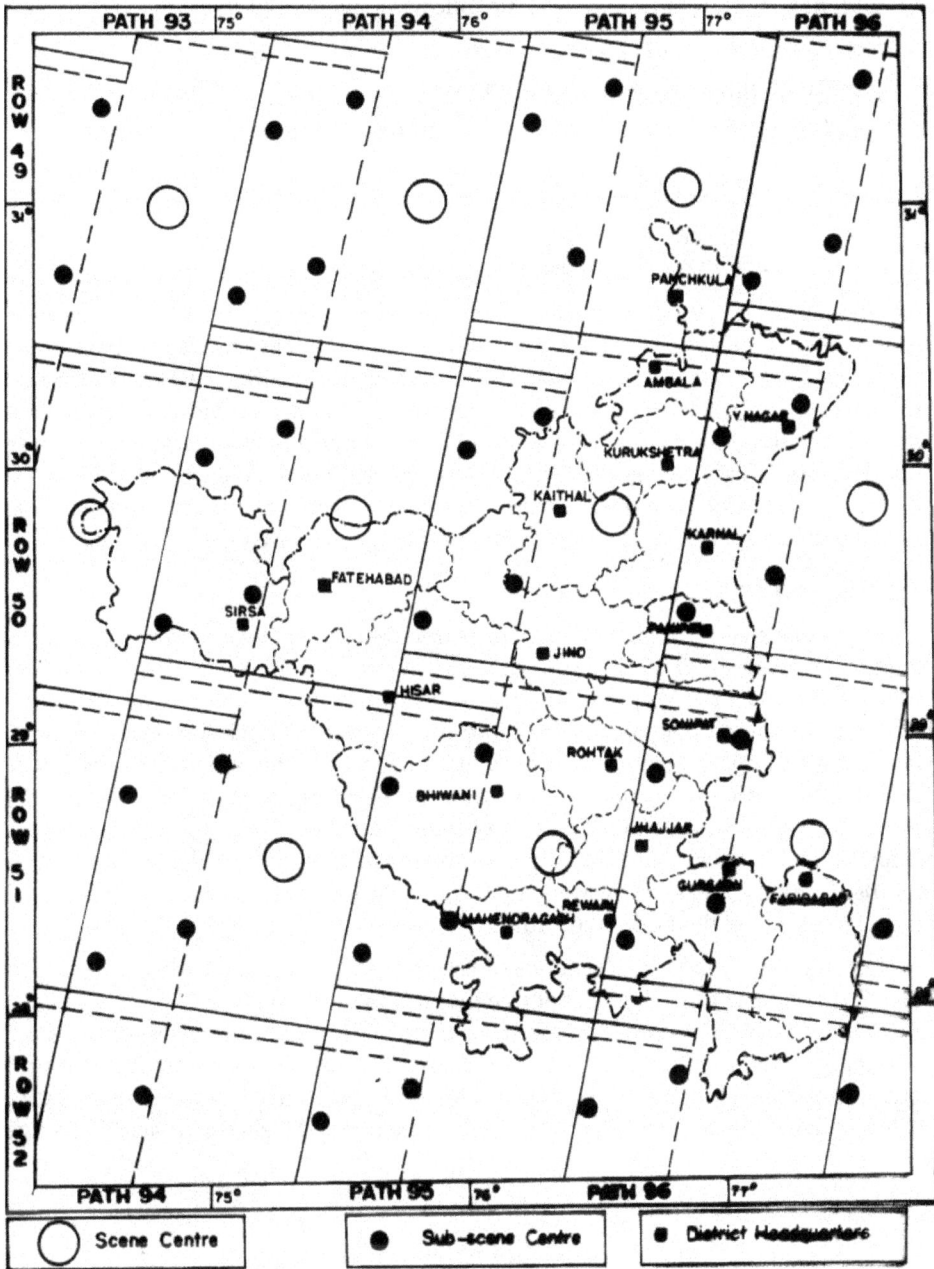

Figure 5.1: Satellite coverage of Haryana state.

Objectives

The study was undertaken with the following major objectives:

☆ Mapping of flood affected areas

☆ To identify various causes of floods

☆ Suggestion of suitable measures to minimize the impact of floods

Data Base and Methodology

IRS 1A/B, LISS II (FCC) diapositive of 15,16 and 17 September 1995 and LANDSAT TM (FCC) film positives of September 24, 1995 were used in the mapping.

Base maps of the entire flood affected areas of the state *i.e.* 13 districts were prepared using Survey of India (SOI) toposheets on 1: 50,000 scale. These base maps were prepared on transparent sheets. Satellite data dia-positives were enlarged on Procom-2 enlarger to fit to the scale and were superimposed on the base maps. Two categories of flood affected areas *i.e.* standing water and receded water/wet areas were delineated. These areas under the both categories were calculated and tabulated. The rainfall data from different stations was also collected to see the severity of the rainfall over this period and will help in analyzing various causes of floods.

Results and Discussion

It was observed from the Rainfall data that there were heavy rains over a week ranging from August 29, 1995 to Sept. 6, 1995 ranging from 150 to 260 mm in different parts of Haryana. This coupled with heavy rains in the upper catchment area of the rivers Yamuna and Ghaggar resulted into unprecedented flash floods in the state. The intensity of rainfall was so heavy that it had exceeded the infiltration capacity of the soil of the state in various reasons and also the filling capacity of depressional areas. The runoff was also too heavy exceeding the carrying capacity of different channels. Due to poor inland drainage, the excess water could not drain out for even months. In many areas, the water table was near the surface and even the small rainfall has created the flood like situation.

District wise flood affected areas of the state in descending order are given in Table 5.1. It is clear from the table above that Rohtak was the worst affected district followed by Bhiwani, Kaithal, Sonipat, Jind and Hisar. Other districts of the state were marginally affected. It was observed that in most of the areas, Agricultural crops, human and cattle population along with settlements and Infrastructure facilities were badly affected. A portion of the area flooded is shown in satellite image in Figure 5.2.

Causes of Floods

Floods mainly occurred due to the following major reasons:

Heavy Downpour

The rains of the magnitude 150 to 250 mm in the state coupled with heavy rains in the catchment area of the rivers passing through Haryana state added to the fury of floods.

Table 5.1: District-wise statistics of flood affected areas.

Sl.No.	Districts	Standing Water	Receded Water	Total	Per cent to Total Area
1.	Rohtak	60925	136033	196958	**55.0**
2.	Bhiwani	26592	101640	128232	**28.7**
3	Kaithal	17770	41312	59082	**24.7**
4.	Sonipat	24397	7284	31681	**22.9**
5.	Jind	43900	10931	54831	**20.0**
6.	Hisar	31947	87068	119015	**17.1**
7.	Faridabad	8616	22595	31211	**14.5**
8.	Gurgaon	13534	25539	39073	**14.4**
9.	Karnal	2788	19622	22410	**11.5**
10.	Kurukshetra	1516	6955	8471	**05.4**
11.	Panipat	1165	4685	5850	**04.6**
12.	Rewari	1169	5611	6780	**04.4**
13.	Sirsa	4448	--	4448	**01.0**
	Total	**238767**	**469275**	**708042**	**100**

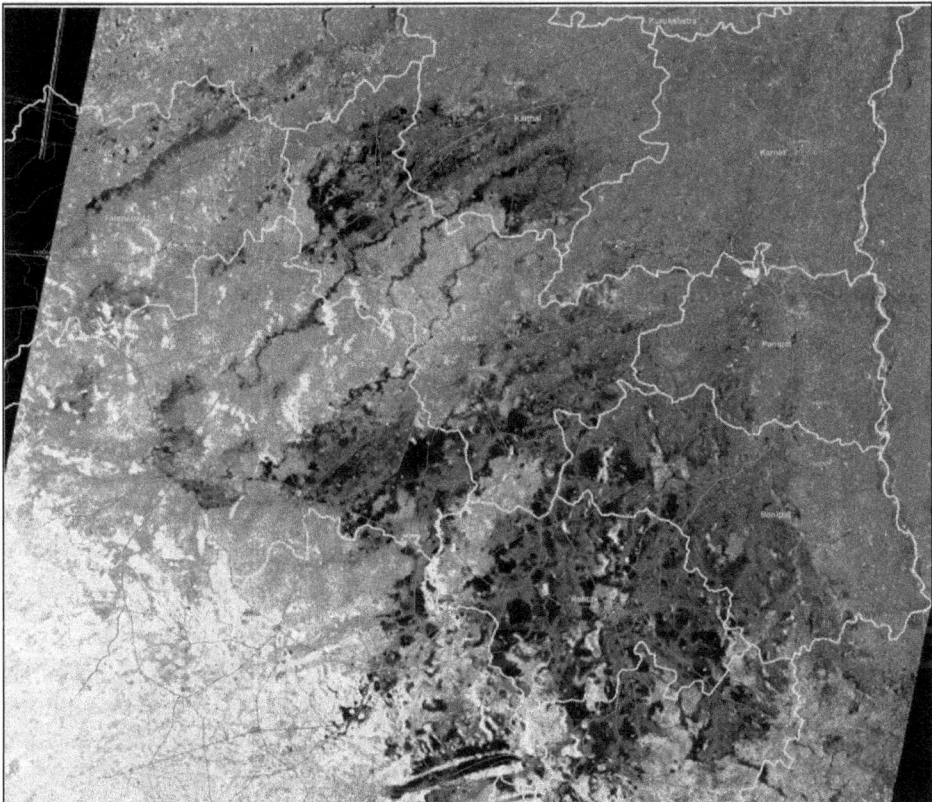

Figure 5.2: Satellite scene of central Haryana (24/9/1995, LANDSAT 5).

Synchronization of Peak Floods

Yamuna and Ghaggar rivers are the prominent rivers in the state. The flow in these rivers was synchronized with the peak flow in tributaries thereby adding to the volume of peak flow. Yamuna affected areas in Karnal, Panipat and Sonipat districts whereas Ghaggar affected Kaithal, Hisar, Jind and Sirsa districts. The surrounding areas along the rivers were badly affected due this. The ephemeral streams like Markanda and its tributaries also show heavy flooding in the areas of Kurukshetra and Kaithal districts.

Poor Inland Drainage

As the general topographic slope of the terrain of the state is towards Southwest, the surface runoff from the northern higher terrain has resulted in the flooding of low lying areas of the state in the central plains. The excess water accumulated in the depressions in Rohtak, Jind, Bhiwani and eastern part of Hisar district resulted into the floods. The water got accumulated in the low lying flat areas and in the absence of proper inland drainage it could not flow out of the area and creating flood like situation for months together. Even the reduced natural water courses have been obstructed by man made activities thereby further resulting in to accumulation of the excess water in the surrounding areas.

The situation was further aggravated by the siltation of drain No. 6 and 8. The course of these drains has reduced to such an extent that even moderate runoff could not be accommodated by these channels. Sonipat district was the victim of flooding due to reduced water carrying capacity of these drains and the excess water got accumulated in the surrounding areas. Flooding in the low-lying areas along the Yamuna River was the result of heavy downpour in the catchment areas.

Canal Breach

The breaching of Petwar minor, Branch of Western Yamuna Canal nears the villages of Kumbha and Thurana and Hisar Distributary near village kheri Gangan and Western Yamuna Canal (Barwala branch) near village Madlauda, north of Barwala town was the main cause of flooding in eastern part of Hisar district.

Unregulated Flow of Water from Surrounding Areas

The unregulated flow of water from Punjab state has added more water in to Ghaggar and from Rajasthan state to Sahibi, Krishnawati and Dohan rivers have created flood like situation in the areas of Hisar and Mahendergarh districts. This was due to excess precipitation in these areas, which exceeded the infiltration capacity of the soils and flowed in the form of surface runoff to meet these seasonal rivers.

Reduction in Effective Water Way of Natural Streams

Many river courses, which were quite wide, have been reduced to almost nallas due to human interferences thereby reducing the carrying capacity of these courses. The excess water has created flood like situation in surrounding areas. These include Ghaggar, Ghaggar, Krishnawati, Dohan and Sahibi rivers course in addition to many other small streams.

Submergence around Depressions

Local depressions in many areas were flooded due to heavy rains during a period of one week resulting into submergence of the areas around these depressions. This was observed mostly in Bhiwani, Gurgaon and Rewari districts. Due to sufficient elevation difference, the natural drainage of floodwater in Gurgaon district was quicker. This was further coupled with the good percolation conditions of the soil due to its sandy nature and deep water table conditions.

Improper De-siltation of Canals

There was growth of weeds in many of the canals, which was not dredged before the onset of monsoon season. This has led to reduction in the carrying capacity of these canals. These canals were not able to accommodate the floodwater and resulted into the overflowing from the banks and breaching of embankments.

Remedial Measures

Floods are naturally occurring event and it is therefore impossible to completely avoid floods. But the fury of floods can be reduced by taking various preventive measures and identifying various reasons from the floods. After studying various causes, it is bit easier to understand about the possibility of various control measures. In the present study, the following control measures are recommended for the area:

1. Warning and Evacuation

The first and foremost requirement for the state is to have a strong warning and communication network to help in effective evacuation. This will involve providing the real time information about the flood in advance and also identification of various sites, which are topographically higher in these areas. This can be done by using integrated RS and GIS methods for terrain modeling and flood forecasting by simulating various situation and identifying probable areas of submergence for various situations. This component is still lacking in the state.

2. Creation of Disaster Management Board

The state does not have any such authority of organization, which can centrally control flood situations. There should be proper mechanism for mitigation the damages due to floods.

3. Strengthening of River/Canal Banks

It is very important to identify vulnerable sites along the river courses to strengthen these before the onset of monsoon. Different types of control works can be carried out for the purpose. The earthen banks be provided with vegetation cover to reduce soils erosion and breaching.

4. Periodic De-silting and Cleaning

To maintain the carrying capacity of the rivers, canals and reservoirs it is necessary to carry out desilting operations periodically before on the onset of the monsoon. This will not only help in maintaining the effective water carrying capacity of canals but will also help in increased capacity of the reservoirs storage.

5. Proper Drainage for Poor Inland Drained Areas

It is necessary to provide proper drainage conditions in the areas that are having poor inland drainage. It is necessary to join the depressional areas all along to create surface drains to carry out flushing operations during monsoon season. This will help in reducing the fury of floods in these areas.

6. Construction of Ring Bunds

The settlements can b e protected by constructing ring bunds around these. In the event of excess rains the water can be pumped out and human and cattle population along with the property assets can be saved in this manner.

There is no single solution for mitigating the fury of floods but a combination of various preventive measures will help a lot in this direction. It is observed that in the areas where floods are frequent, people are well trained in taking effective steps and thereby loss is less in comparison to the areas where floods are observed once in a while. It is therefore necessary to train the people in various simple methods to face this type of challenge. Active participation of the NGO's and formation of State Disaster Management Board will be of great help.

Conclusions

The study demonstrates the role of Remote Sensing techniques for assessment and mitigation of natural disasters like floods. Being a natural phenomenon, it is impossible to completely stop the floods but these can be mitigated by taking various preventive measures in this direction. One of the important factors is the creation of proper warning and communications system to avoid the panic among the people. The creation of State Disaster Management Board in the state will also help in creating a database and demarcation of various risk zones in the state. RS and GIS will be of great help in this direction.

References

Chaudhary, B.S.; Arya, V.S.; Beniwal, A.; Babu, T.P. and Ruhal, D.S. (1999). Satellite Applications in monitoring flood hazards in Haryana, India- A Case study. Operational Remote Sensing for sustainable development. A.A. Balkema/ Rotterdam/Brookfield/1999. Edt. G.J.A. Nieuwenhius, R.A.Vaughan and M.Molenar. p. 281-284.

De Brouwer, J.A.M. (1994). Flood study in the Meghna Dhonagoda polder, Bangladesh. Proceedings 15th Asian Remote Sensing Conference. Vol II. P. C-7-1-C-7-6.

Florenzano, T.G. (1989). Flood management using remote sensing data- Brazilian experience. Space and Flood management, I.A.F. 40th Congress, Malaga, Spain. p.31-46

Ramamoorthy, A.S. (1989). Flood management-applications of satellite remote sensing. Space and Flood management. I.A.F. 40th Congress, Malaga, Spain. pp.9-30.

Rao, U.R. (1989): Space and Flood management. *Ibid.* pp. 1-7.

Nageshwar Rao P.P. and Jayraman,V. (1992). Satellite Remote Sensing in Disaster Managemnt. Natural Resources Management- A new perspective. NNRMS, Department of Space, Bangalore. pp. 211-233.

Navalgund R.R. (2006): Geoinformatics in disaster monitoring and mitigation. ISG Newsletter Vol.12, No.1 and 2. pp. 4-23.

Sharma, P.K. (1989): Extent of flooding in Punjab during 1988. *NNRMS Bulletin* Vol. 12. p. 35.

6

Drought Occurrences and Management in Kenya

Jones F. AGWATA

Introduction

Mankind has always been faced with extreme hydro-climatic disasters and problems related to either too little or too much water. Water related disasters have been more devastating as far as deaths, suffering and economic damages are concerned, compared to other natural hazards such as earthquakes and volcanoes. Besides the destructive direct effects, extreme hydro-climatic events have often been followed by secondary, indirect calamities such as famine, epidemics and fire.

Despite continued advances in research, mankind is still very much vulnerable to extreme hydro-climatic and natural hazards, which despite heavy expenditures on both structural and non-structural measures for their control, continue to cause untold suffering and huge economic losses in both developed and developing countries. For instance, over the last ten years, the world has witnessed a 68 per cent increase in the number of disasters that have affected more than 2.5 billion people, killing 478 100 and caused economic losses estimated at about US$ 690 billion (Briceno, 2005). The impacts are far deeper and more serious in terms of human losses and societal effects especially in developing countries. Understanding the characteristics of these events is crucial not only for their management but also for allocation of necessary resources and the design and management of various resource management systems.

Drought is one example of an extreme hydro-climatic event or hazard that recurs and is a normal part of climate. It differs from other natural hazards like floods, earthquakes, storms and landslides in that its effects accumulate slowly and lingers for long even after the event is over and its onset and termination dates are difficult to determine. Besides, it lacks a universal definition, its impacts spread over very large areas making the quantification of the impacts much more difficult and it differs in duration, frequency, magnitude and severity (Smakhtin, 2001).

Understanding and Defining Drought

Drought is considered to be a normal, recurrent feature of climate and it occurs in different climatic zones although its characteristics vary significantly from one region to another. It differs from aridity, which is restricted to low rainfall regions and is a permanent feature of climate. Although it has various definitions, it originates from a deficiency of precipitation over an extended period of time, usually a season or more. This deficiency results in a water shortage for some activity, group or environmental sector.

It lacks a common, precise and universally accepted definition. This is due to the fact that it means different things to different people and is region, impact and depends on the discipline. Consequently, the available definitions are either conceptual, operational or discipline based. Conceptual definitions are formulated in general terms and help people to understand the concept of drought. The definitions are important in establishing drought policy as national drought policies incorporate understanding normal climate variability into definitions of drought. The definitions also define boundaries of drought and are generic in their description.

Operational definitions identify the onset, termination and the degree of severity of drought occurrences and are useful in analysing frequency, severity and duration of drought. The definitions specify the degree of departure from and average of precipitation or some other climatic variable over some time period by comparing the current situation to the historical average, often based on a 30-year period of record. The threshold identified as the beginning of a drought is usually established somewhat arbitrarily, rather than on the basis of the precise relationship with specific impacts. Operational definitions can be used to analyze drought frequency, severity and duration for a given historical period. Such definitions, however, require data on hourly, monthly, or other time scales depending on the type of definition being applied. The development of drought climatology of a region provides a greater understanding of its characteristic and the probability of recurrence at various levels of severity; information that is extremely beneficial in the development of response and mitigation strategies and preparedness plans.

According to discipline, drought may be viewed as meteorological, agricultural, hydrological, urban and socioeconomic in type. Meteorological drought is defined usually on the basis of the degree of dryness and the duration of the dry period and definitions have to be considered region specific as the atmospheric conditions that result in deficiencies of precipitation vary from region to region. For example, some definitions identify periods of drought on the basis of the number of days with precipitation less than some specified threshold. Agricultural drought links various characteristics of meteorological or hydrological drought to agricultural impacts, focusing on precipitation shortages, differences between actual and potential evapotranspiration, soil water deficits, and reduced ground water and reservoir levels. Since plant water demand depends on prevailing weather conditions, biological characteristics of the plant, its state of growth, and the physical and biological properties of the soil, a good definition of agricultural drought should be able to account for the variable susceptibility of crops during different stages of crop developments. This is because deficient soil moisture at planting may hinder germination, leading to low plant populations and eventually the final yield. However, if top-level moisture is sufficient for early growth requirements, deficiencies in subsoil moisture at this early stage may not affect final yield if subsoil moisture is replenished as the growing season progresses.

Hydrological drought is associated with the effects of precipitation shortfalls on surface or subsurface water supply in the form of stream flow, reservoir and lake levels and ground water. The frequency and severity of hydrological drought is often defined on a watershed or river basin scale and even though all droughts originate from a deficiency of precipitation, hydrologists are more concerned with how this deficiency plays out through the hydrologic system. This drought is usually out of phase with or lags the occurrence of meteorological and agricultural droughts. It takes longer for precipitation deficiencies to show up in components of the hydrological system such as soil moisture, stream, stream flow, and ground water and reservoir levels. Consequently, these impacts are out of phase with impacts in other economic sectors. The drought also implies high socio-economic and human losses particularly where rivers act as water supply systems or as inflows to hydropower, and faunal habitats. Hydrological drought has implications in the planning and design of water supply systems, allocation of waste loads, design of reservoir storage and maintenance of quantity and quality of water for irrigation, recreation and wildlife conservation.

Whilst urban drought is said to occur when actual water supply is insufficient to meet municipal water demand under the normal operation of the water supply system (Rossi *et al.*, 1992), Socio-economic drought occurs when the supply and demand of some economic good is related to elements of meteorological, hydrological, and agricultural drought. It differs from the previously mentioned drought types since its occurrence depends on the time and space processes of supply and demand to identify or classify droughts. Supply of many economic goods, such as water, forage, food grains, fish, and hydroelectric power depend on weather. In most instances, the demand for economic goods increases due to increasing population and per capita consumption. Supply may also increase because of improved production efficiency, technology, or the construction of reservoirs that increase the surface water storage capacity.

A detailed account of the various perspectives and manifestations of drought has been provided by Wilhite and Glantz (1985, 1987), UNEP (1992), Kundzewicz *et al* (1993) and WMO (1997), among others.

Occurrence of Drought in Kenya

Drought should be considered relative to some long-term average condition of the balance between precipitation and evapo-transpiration in a particular area, a condition often perceived as "normal". It can also be related to the timing and effectiveness of the rains although other climatic factors such as temperature, wind and relative humidity are often associated with it in many regions of the world and can significantly aggravate its severity. Recent droughts in both developing and developed countries and the resulting economic and environmental impacts and personal hardships have underscored the vulnerability of all societies to this "natural" hazard and necessitated that appropriate strategies should be put in place to manage it in such away as to reduce various impacts associated with it.

Drought is particularly prevalent in the arid and semi arid lands (ASALs) of most countries including Kenya. It is an insidious hazard of nature and should not be viewed as merely a physical phenomenon or natural event since its impacts arise due to the interplay between the event and the demand that is placed on water supply by people. The ASALs in Kenya cover about 88 per cent of the country's land surface area and are characterized by strong variations in rainfall, high temperatures and high evaporation rates and support 30 per cent of the country's total human population, over 50 per cent of the livestock and between 80-90 per cent of wildlife and yet these areas are characterized by unpredictable rainfall patterns, high evaporation rates, low organic matter levels and poor infrastructure. The arid parts receive less than 400 mm of annual rainfall while the semi arid areas receive rainfall of between 400 and 1 000 mm annually (NEMA, 2005 and Nikundiwe and Kabigumila, 2006). These areas are prone to harsh weather conditions rendering the communities within the ecosystems prone to frequent drought episodes (Republic of Kenya, 2004).

Kenya experiences cyclic drought episodes with the major ones coming every ten years and the minor ones almost after every three to four years. The 2004 drought is a replica of the previous cycle of severe droughts that affected the country as experienced in 1974, 1984 and 1994. Drought occurrence in the country has been recurrent with spatial and temporal characteristics as shown in Table 6.1.

Table 6.1: Recent drought incidences in Kenya.

Year	Area of Coverage	Number of People Affected
1975	Widespread	16 000
1977	Widespread	20 000
1980	Widespread	40 000
1983/1984	Widespread	200 000
1991/1992	Most Arid and Semi Arid Districts	1.5 million
1995/1996	Widespread	1.5 million
1999/2000	Widespread	4.7 million
2004	Widespread	2-3 million

Source: UNDP, 2004.

The country has in the past recorded food deficits due to drought resulting from rainfall deficits in 1896-1900, 1925, 1928, 1933-34, 1937, 1939, 1942-44 1947, 1951, 1952-55, 1957-58, 1972, 1973-74, 1984-85 and 1999-2000. The 1983-84 and 1999-2000 drought episodes are recorded as the most severe that led to losses of human life, livestock and heavy government expenditure to facilitate appropriate responses (UNDP *et al.*, 2002; UNEP and GOK, 2002).

Several factors are known to contribute to the severity of drought in the country. These factors include inadequate water storage capacity, increased destruction of forests due to charcoal burning, clearance of forests for agriculture and frequent forest fires. Others causative factors include poor management of catchment areas, cultivation on stream banks and steep slopes causing erosion of soil that silts up dams and pans that are used as dry weather water reservoirs, lack of policy of managing water and drought and inadequate distribution of water resources. Several factors are known to increase the vulnerability of communities in the ASALs which together form a complex web of events that cumulatively act to reduce the ability of the people to cope with drought. Vulnerability would be increased as a result of unfavorable climate or lack of irrigation equipment/resources, lack of knowledge and skills, preferences by farmers, unclear or absent drought warnings, lack of research and coordination and government bureaucracy.

Several social, economic, physical and environmental factors increase vulnerability in the country's drought prone zones. Examples of social issues include indigenous beliefs, traditions, insecurity and different ways of coping with the drought. Others include literacy levels which affects the understanding of drought as a hazard. The economic factors that increase the vulnerability of communities to the vagaries of drought include rise in food prices, depletion of food reserves, deterioration of health due to lack of food and clean water. Poor infrastructure such as impassable roads, poor telecommunication lines and inaccessibility of some regions hampers the transportation of food and also action in terms of response to distress calls, poor publicity and inability to air the plight of the people. Drought has been increasing in severity in the country in the past four decades and this has been associated with environmental degradation due to increasing urbanization, development, extension of agricultural land into forested areas and the logging of trees for charcoal burning. The result of all these land use changes is an increase in wild fires as a result of high temperatures during the drought season. These fires have played a big role in environmental degradation rendering the ASALs even more prone to drought conditions as a result of vegetation depletion (UNEP and GOK, 2002 and UNDP, 2004).

Impacts of Drought

The effects of drought are diverse and include reduced agricultural production, plant and animal damage, air and water pollution, forest fires and increased demand for water and the ensuing conflicts among various water dependent activities. Other effects include disruption in the performance of water supply systems, decline in hydropower production, dilution capability of rivers and depletion of groundwater aquifers. The occurrence of drought may also lead to stream flow deficits and water level fluctuations in several rivers, reservoirs and groundwater aquifers in water basins causing undue pressure on the water resources (UNEP/GOK, 2002). Droughts have been recorded to occur between three to ten year periods and the drought of 2004 caused severe water shortage that affected the agricultural and industrial sectors resulting in acute human and animal suffering as well as degradation of environmental resources. The drought of 1999/2000 affected 4.7 million people causing acute and chronic malnutrition rates due to diseases, which increases mortality rates. Other effects include loss of livelihoods leading to over reliance and unsustainable utilization of natural resources (NEMA, 2005).

The differences in the drought types may be seen in the sequence of impacts associated with various drought types. When drought begins, the agricultural sector is usually the first to be affected due to its heavy dependence on stored soil water. Soil water can be rapidly depleted during extended dry periods. If precipitation deficiencies continue, the people dependent on other sources of water will begin to feel the effects of the shortage. Those who rely on surface water from reservoirs and lakes and subsurface water, for example, are usually the last to be affected. A short-term drought that persists for 3 to 6 months may have little impact on these sectors, depending on the characteristics of the hydrologic system and water use requirements. When precipitation returns to normal and meteorological drought conditions have abated, the sequence is repeated for the recovery of surface and subsurface water supplies. Soil water reserves are replenished first, followed by stream flow, reservoirs and lakes, and ground water. Drought impacts may diminish rapidly in the agricultural sector because of its reliance on soil water, but linger for months or even years in other sectors dependent on stored surface or subsurface supplies. Ground water users, often the last to be affected by the drought during its onset, may be last to experience a return to normal water levels. This length of the recovery period is a function of the intensity of the drought, its duration and the quantity of precipitation received as the episode terminates.

Drought impacts in Kenya may be classified into both long term and short term economic and social impacts. The short term socioeconomic impacts include migration and displacement of people into areas with relief foods, malnutrition leading to ill health, price hikes for commodities such as cereals, and other food products whilst livestock prices go lower, lack of social amenities like water, food and sanitation services, livestock diseases and low yields or no yields from agricultural activities as a result of low moisture content in the soil. The long term impacts include loss of livelihood, and paralyzed economic activities, poor health, deaths of children and the old, increased poverty, overall dependence on relief supplies from Government and other agencies, increased conflicts due to diminished water and food resources, politics and invasion of surrounding communities. The areas affected by drought have experienced reduced water levels leading to human, livestock and wildlife populations' concentration near available water points causing degradation, loss of livestock and high incidences of disease outbreaks due to poor hygiene and sanitation practices as well as poor water quality and inadequate quantities.

Drought Management

Due to frequent drought occurrences in the ASALs of Kenya, the government has put in place programmes designed to minimize risks associated with drought and this has helped to put in place instruments for early warning, contingency plans and mitigation strategies and mechanisms. The drought recovery initiatives include drought management project (DMP), Drought Preparedness, Intervention and Recovery Project (DPIRP), Emergency Drought Recovery Project (EDRP) and the Arid Lands Resource Management Project (ALRMP).

Besides the various mechanisms that have been put in place to manage drought, various mitigation measures should be instituted to cope with drought. These measures would include:

☆ Increasing food aid and non-food interventions to assist with emergence health kits and critical immunization equipment to drought affected areas.

☆ Improving reliable access to safe water, hygiene and sanitation facilities for drought affected communities.

☆ Strengthening the Early Warning Systems to enhance effective drought monitoring systems in the affected areas.

☆ Adoption of participatory community drought management initiatives through strategic drought management and contingency plans.

☆ Setting up of drought mitigation and development priorities through participatory appraisal mechanisms.

☆ Ensuring access to water in drought affected areas with the aim of facilitating conflict resolution, supporting water tinkering to remote communities and provision of emergency repairs of boreholes and desiltation of dams and reservoirs.

☆ Provision of contingency funds to equipping selected pasture areas with temporary water supplies replenished by tinkering as grazing reserves.

☆ Establishment of alternative water sources such as boreholes, water pans, shallow wells for both livestock and human well being and health.

Other suggestions for drought management would include public education focusing on drought occurrence, impacts and mitigation mechanisms in various drought-prone areas of the country. Mechanisms of maintaining holdover storages, digging of deeper wells, lowering intakes to groundwater supplies and adopting various water conservation activities such as reforestation, catchment rehabilitation should also be instituted. In addition, sustainable management and control of surface run off should be promoted through strip cropping, terracing, storage of water in farm ponds, back of check dams and reservoirs so as to provide for domestic and agricultural activities in ASAL areas. Due to increasing demand for water for various uses in different parts where drought is frequent, it is necessary to limit or adjust different economic development activities to available water. The various zones prone to drought occurrences should be used to guide future development policies and endeavours in this regard. These strategies are particularly critical in the case of hydrological drought in the critical water basins in the country (Agwata, 2005).

A comprehensive drought management policy should be instituted to provide the basis for integration of drought management issues into national development programmes. A suitable mechanism is needed to bring together all the stakeholders such as government, nongovernmental and international organizations, institutes of higher learning and research and various local communities to be actively involved in drought management activities in the arid and semi arid parts of the country. An early warning and national monitoring system will also need to be established to ensure continuous assessment of drought impact in various regions of the basin. The key elements of such a policy especially suited for pastoral communities would include an appropriate institutional structure, early warning and preparedness, rapid reaction, reduction of livestock mortalities, livestock interventions, participation of the affected communities in decision making and coordination and management of drought mitigation (Swift, 2000; Morton *et al.*, 2002 and Swift *et al* 2004).

The policy should also address several other issues including identifying and setting aside land marks for meeting urgent shelter requirements for people affected by drought, delineation of high risk drought areas, developing procedures for sustainable development of drought prone areas and promoting research on use of indigenous knowledge in the management of drought and drought resistant traditional food crops. It should also assist in promoting drought mitigation mechanisms at the community level, establishing appropriate data banks and information systems on drought and ensure timely dissemination to affected parties to enhance community awareness and preparedness, provide for capacity building on drought management through implementation of land use development plans, undertake monitoring and evaluation as a means of response and using past drought audits to take preventive measures that minimize negative impacts and mainstream comprehensive drought emergency management and response plans and provide for an integrated approach in dealing with the causes and other characteristics of drought (Mango *et al.*, 2007).

Other areas that the policy should address include the following

☆ Integration of strategies for poverty eradication into efforts to mitigate drought.

☆ Promotion of cooperation among affected communities in environmental protection and conservation of land and water resources as they relate to drought occurrence.

☆ Emphasize implementation of preventive measures against degradation of land.

☆ Enhancing capabilities to provide early warning for drought occurrence

☆ Strengthening drought preparedness and management including contingency plans that take into account seasonal and interannual climate variability.

☆ Establishing and strengthening food security systems as well as alternative livelihood projects that could provide incomes in drought prone areas.

☆ Developing sustainable irrigation projects for the raising of crops and livestock.

☆ Conservation of resources through sustainable management of agricultural and pastoral land, vegetation cover, wildlife, forests, water resources and biological diversity.

☆ Development and use of alternative energy sources such as wind, solar and biogas through adaptation transfer and acquisition of appropriate technologies to alleviate the pressure on fragile resources.

The policy should ensure appropriate responses for minimizing various impacts triggered by drought occurrence. Some of the objectives of the policy would include maintenance of purchasing power of the affected and vulnerable communities, ensuring availability of cereals and grains, provision of adequate human nutrition and health care, sustaining enrollment of pupils in schools through the provision of food, and ensuring availability of water for both human and livestock use. To be able to achieve the objectives, an institutional capacity should be in place to generate accurate early warning and monitoring data at community, district and national levels, intervene in a timely manner during drought emergencies, manage responses and rehabilitation activities during the drought episode and ensure availability of adaptable contingency plans at district level to ensure implementation of appropriate drought responses (Swift *et al.*, 2004).

Drought management in the country faces several challenges. Some of these challenges include expensive training facilities, lack of appropriate early warning systems, mapping equipment, non existent linkages between various institutions and other global disaster monitoring centers and weak research collaboration between social and physical scientists (Nyadawa, 2005). To address these challenges, there is need for a national drought management policy and a drought management coordinating body. Research and information data base on drought management to improve liaison amongst concerned institutions and specialized training and facilities for monitoring and assessment of drought hazards would also go a long way in addressing the challenges facing drought management in the country.

Conclusions and Recommendations

Drought occurrence in Kenya is a frequent cyclic occurrence whose impacts are enormous to the socioeconomic development of the country. Since drought can not be prevented from occurring, mechanisms must be put in place at all levels to manage the impacts related to the drought. This means that drought management and conservation plans should be developed and implemented through the establishment of an integrated early warning system with a drought communications system to ensure timely delivery of advance warnings to the public. An appropriate policy should also be established to take care of various aspects of drought management such as early warning, contingency planning, mitigation, relief and rehabilitation. There should also be mechanisms for resilience such as land tenure, appropriate institutions, and support for marketing services, infrastructure and security, among others. Various districts that are prone to drought should develop their drought management plans for purposes of managing the drought as and when it does occur.

It is recommended that to manage drought in an integrated way, there is need for

☆ Appropriate capacity building and establishment of an appropriate infrastructure.

☆ Data management on issues related to drought.

☆ Proper systems for timely information sharing and communication.

☆ Research and monitoring of patterns of adaptation and traditional coping strategies to the drought hazard.

☆ Development of best practices on natural resources use.

☆ Education, training and public awareness on issues related to drought occurrence, impacts and vulnerability assessments.

☆ Financial allocations for drought management activities and sensitization at the community level.

References

Agwata, J. F. 2005, *The Characteristics of Hydrological Drought in the Upper Parts of the Tana Basin, Kenya. Unpublished PhD thesis,* Kenyatta University, Nairobi, Kenya.

Brecino, P. 2005. All are responsible for disaster reduction. In: *Environment and Poverty Times,* January 2005, UNEP-ARENDALL Publication.

Kundzewicz, Z. W., Rosbjerg, D., Simonovic, S. P. and Takeuchi, K. 1993, Extreme Hydrological Events: Precipitation, Floods and Droughts. *Proceedings of the International Conference held at Yokohama, Japan. IAHS Publication No. 213,* pp. 1-7.

Mango, N, Kirui, A. I. And Yitambe, A. 2007, Status of Disaster risk Management in Kenya. In: Fuchaka, W., Otor, S., Olukoye, G and Mugendi, D. (Editors), *Environment and Sustainable Development, a Guide for Higher Education in Kenya, Vol II,* School of Environmental Studies and Human Sciences, Kenyatta University, Nairobi, Kenya. pp. 158-178.

Morton, J., Barton, D., Collinson, C. and Heath, B. 2002, *Comparing Drought Mitigation Interventions in the Pastoral Livestock Sector,* NRI Report.

National Management Authority, NEMA 2005, *State of the Environment Report, Kenya 2004: Land use and Environment,* NEMA, Nairobi. pp. 72-81, 138-139.

Nikundiwe, A. M. and Kabigumila, J. D. L. (Editors), 2006, *Drylands Ecosystems: Challenges and Opportunities for Sustainable Natural Resources Management, Research Programme on Sustainable Use of Dry Lands Biodiversity,* University of Dar es Salaam, Tanzania.

Nyadawa, M. O. 2005, Evaluation of Civil engineering Training in Respect to Disaster Preparedness in Kenya. *Proceedings of the 2005 Scientific, technological and industrialization Conference, 27th-28th October, 2005, Nairobi, Kenya.* pp. 353-357, ISBN 9966-923-28-4.

Ogola, J. S., Abira, M. A. and Awuor, V. O. (Editors) 1997, *Potential impacts of Climate Change in Kenya, Climate Network Africa, Nairobi. Chapter 9: Climate Change: Drought and Desertification Kenya.*

Republic of Kenya, 2004, *National Policy for the Sustainable Development of the Arid and Semi Arid Lands of Kenya*, Nairobi, Kenya.

Swift, J., Barton, D. and Morton, J. 2004, *Drought management for pastoral livelihoods-policy guidelines for Kenya*. NRI Report.

Swift, J. 2000, *The Institutional Structure for Drought Management in Kenya*. Acacia Consultants, Nairobi, Kenya.

UNEP, 1992, *Preparing for Drought: A Guide Book for Developing Countries*. UNEP, Nairobi.

UNEP and GOK, 2002, *Devastating Drought in Kenya, environmental impacts and responses*. UNEP, Nairobi.

UNDP, 2004, *Kenya Natural Disaster Profile*, UNDP, Nairobi, Kenya 26 p.

UNDP, WMO, GOK, IGAD and DMCN, 2002, *Factoring Weather and Climate Information and Products into Disaster Management Policy: A Contribution to Strategies for Disaster reduction in Kenya*, Nairobi, Kenya.

Wilhite, D. A. and Glantz, M. H. 1985, Understanding the Drought Phenomenon: The Role of Definitions. *Water International*, **10:** 111-120.

Wilhite, D. A. and Glantz, M. H. 1987, Understanding the drought phenomenon: The role of definitions. In: D. A. Wilhite and W. E. Eastering (Editors), *Planning for drought: Toward a Reduction of Societal Vulnerability*. Westview Press, Boulder, Colorado. pp. 11-27.

WMO, 1997, *Drought Preparedness and Management for West African Countries*. WMO, Geneva, Switzerland.

7

An Interdisciplinary Approach to Understanding Landslides and Risk Management: A case study from Earthquake Affected Kashmir

K. Sudmeier-Rieux, R.A. Qureshi, P. Peduzzi, M.J. Jaboyedoff,
A. Breguet, J. Dubois, R. Jaubert, and M.A. Cheema

Introduction

The October 8, 2005 earthquake (EQ), measuring 7.6 on the Richter scale caused an estimated 73,000 casualties and triggered an estimated 1,000 landslides affecting a large number of communities in surrounding steep mountain valleys. Landslides remain a great threat to communities, especially during heavy rainfall and July/August monsoon rains (GSP 2007a, Petley *et al.*, 2006). The goal of this chapter is to describe an interdisciplinary methodology and findings from a one-year study of landslides and underlying factors rooted in land use and land use strategies that affect the vulnerability of communities in Neelum Valley, northeast of Muzaffarabad, in earthquake-affected Kashmir. Main findings of data collected on 100 landslides, 17 crack zones and 3 flood areas demonstrate a predominance of grazing, deforestation and road construction as "preparatory factors" (Crozier 1986). The economic damage from the landslides was considerable, estimated at 72 million PKR ($US1.2 million), not including damages to the power supply, estimated at 238 million PKR ($US3.9 million), of which half was likely due to landslides. Findings of the socio-economic survey conclude that for the two villages surveyed, risk perception of future

Figure 7.1: Neelum valley road.
Karen Sudmeier-Rieux, UNIL 2007.

landslides is high Most respondents are aware of the danger posed by cracks and households are avoiding risk by reconstructing in safer places within their village. With a few exceptions, respondents had not clearly made the link between deforestation and landslides and many requested assistance and information on how to mitigate landslides. This interdisciplinary approach to assessing landslides offers policy makers a more holistic picture of the underlying causes of landslides and an improved basis for designing a sustainable disaster risk reduction strategy.

The research project was undertaken by a consortium of partners lead by the World Conservation Union (IUCN) Ecosystem Management Programme, IUCN-Pakistan, the University of Lausanne, Institute for Geomatics and Risk Analysis, UNEP-GRID/EUROPE, Division of Early Warning and Assessment and the Graduate Institute for Development Studies, Geneva, funded by the Geneva International Academic Network (GIAN).

Rationale for Study

An Interdisciplinary Approach to Understanding Landslides

The number and frequency of "natural disasters" is affecting greater numbers of vulnerable populations, especially in coastal- and mountain areas (ISDR 2004). Disasters happen to people, especially women who are put at risk due to their vulnerability and are more often a product of climate change (Cannon, 1994).

Therefore, the use of "natural" to describe disasters is a misnomer (Hewitt 1983, Abramovitz *et al.*, 2002). Another facet, which is beginning to receive greater attention in the literature on hazard events is the environmental dimension, *i.e.* the protective role played by ecosystems in abating the full impact of hazard events (ISDR 2007, Sudmeier-Rieux *et al.*, 2006, Abramovitz *et al.*, 2002). The Millenium Ecosystem Assessment, an international study of the world's ecosystems has brought attention to ecosystem services and their value in supporting human livelihoods (MEA 2006). Organizations such as the World Conservation Union (IUCN) and the U.N. Environment Programme (UNEP) are emphasizing the protective roles played by ecosystems for disaster risk reduction, such as coastal mangroves, sand dunes, coral reefs, riverine estuaries, inland wetlands, and forests on steep slopes.

In the Kashmir EQ aftermath, the devastation caused by the massive and pervasive landslides can be attributed to many factors: population pressure leading to deforestation and poor road building undercutting already fragile slopes, as the consequence of poor governance and development. Disasters create the need for immediate, short-term reaction but effective disaster management requires long-term, systematic solutions (Smith 2004). Any effective strategy for mitigating this complex cause and effect between ecosystems and social systems must be interdisciplinary by nature.

Landslides and the Importance of Vegetative Cover

Landslides can be considered a symptom of fragility, either natural or human-induced. A small seismic shock to a sensitive system can cause a landslide, whereas a system with higher buffering capacity may sustain little reaction to seismic shock (Hufschmidt *et al.*, 2005). In the case of Kashmir, a highly sensitive system received a great shock, resulting in massive landslide damage. Causes of landslide susceptibility are multiple: weak geological structures (limestone, silt and clays), morphological, tectonic uplift, physical (intense rainfall, earthquake), and anthropogenic (excavation of slope toe for roads, loading of slope due to irrigation, deforestation, irrigation) (Cruden and Varnes 1996). Anthropogenic factors are considered "preparatory factors", whereas rainfall or earthquakes are "triggering factors" (Crozier 1986). Rainfall can actually be considered both: it contributes to slope instability and it triggers landslides.

Deep root systems are believed to play a major role in holding soil in place on steep slopes, thus the importance of maintaining vegetative cover. Even under conditions of high rainfall, steep forested slopes are generally stable (Alexander 2005). This stability is assured if forest cover is continuously maintained, but indiscriminate removal of forest followed by high rainfall, can lead to disastrous results (Pryor 1982, Saldivar-Sali and Einstein 2007, Phillips and Marden 2005). Deforestation, if not replaced by other sustainable land use, can contribute to increased frequency and severity of floods and landslides (Zingari and Feibiger 2002, Alexander 2005, Phillips and Marden 2005). Vegetative solutions may not suffice for certain deep-seated landslides, which may require considerable engineering solutions. Such costly techniques may not be readily available to a developing country; however more shallow landslides may be stabilized with locally available materials.

Study Background

Details of Study Area: Neelum Valley, Azad Jammu and Kashmir (AJK)

The study area, lower Neelum Valley, was chosen due to its proximity to Muzaffarabad (pop. 100,000) and its very distinctly forested north-facing left bank, and its south-facing, largely deforested right bank. The left bank has fewer villages, is to a greater extent state-owned, with fewer private and communal lands, or *shamilat*. Neelum is a steep, v-shaped valley with estimated slope range from 35-65° and average width of 15 km. The 2005 EQ epicenter was located here, approximately 19 km from Muzaffarabad (Bulmer *et al.*, 2007). Due to extremely difficult terrain, the valley was blocked for 47 days following massive landslides resulting from the EQ.

The area is densely populated (264 persons per square km) with an average family size of 7.2 members (AJK Planning and Development Department 2005). Approximately 42 per cent of the area is forested, albeit much is degraded, 13 per cent is under cultivation and 42 per cent is considered uncultivable land, used mainly for grazing and the remainder is urbanized (AJK Forest Department 2001). Presently, most of the lowland forests of AJK and Pakistan are badly degraded or entirely destroyed, due to slow upward creep of deforestation for domestic use, grazing, or commercial logging and nationwide, only approximately only 5 per cent of Pakistan is forested (Ahmed and Mahmood 1998).

The average rainfall for the Muzaffarabad region is 1367 mm (average 1995-2000), with 30-60 per cent in the form of snowfall during December- February, usually above 2000m (AJK Planning and Development Department 2005). Monsoon rains occur during July and August, and "cloud bursts" can bring as much as 100mm during one single event, causing significant damage as flash floods or debris flows. With changing climate conditions, more extreme rainfall patterns can be expected throughout the year.

Geologically, the area is situated in the pre-Himalayan zone, along the arc collision zones between the Main Karakorum Thrust (MKT) and the Main Mantle Thrust (MMT) (GSP 2007a). The earthquake was caused by the rupture of the northwest-southwest Muzaffarabad thrust fault (Bulmer *et al.*, 2007). It can be considered as the result of interaction between three tectonic elements *viz.* (a) the Himalayas (b) the Indo-Pakistani Shield and (c) the Salt Range, each of which is moving independently. The thrust zone has a NNW orientation and is largely covered by Murree Formation, which is a mix of sandstone, siltstone and shalestone, followed by the Abbottabad Formation, of dolomitic limestone (GSP 2007a, Schneider 2006). The Murree formation is characterized by impermeability and is susceptible to landslides due to rainfall (GSP 2007b).

Study Goals and Methods

The goal of the study is to strengthen decision-making tools for disaster risk reduction in mountainous areas. By understanding the linkages between land use, land use practices, resources management and disaster risk reduction, development- humanitarian and donor agencies focused on disaster risk reduction can improve their risk reduction programs for mountainous regions.

The objectives are:

1. Identify and analyze the damages and losses caused by the landslides triggered by the 2005 EQ in lower Neelum Valley, AJK.

2. Examine natural and human-induced land use factors related to landslides in lower Neelum Valley, AJK.

3. Estimate to what extent forest cover played a role as natural barriers to landslides.

4. Examine community land use strategies that impact the vulnerability of communities in lower Neelum Valley, AJK.

Research Design

The fundamental research question of the study is *what role did land use factors, especially: vegetative cover, roads, terraces, and ownership play in the occurrence of landslides in the study area?* Our hypothesis is that land use, especially vegetative cover and road construction will significantly impact the occurrence of landslides.

A second research question relates to *the usefulness of socio-economic data: land use, ownership, risk perceptions in addition to geological information for improving decision-making tools for disaster risk reduction in mountainous regions.*

The interdisciplinary study has four main components:

1. Assessment of information needs, existing information, data and literature.

2. Landslide susceptibility model and map, *"Modelling probability of landslides occurrence and potential link with deforestation in North Pakistan"* was developed by Mr. P. Peduzzi (UNEP-GRID/EUROPE), assisted by intern/student Mr. R. Klaus (Klaus, 2007).

3. Landslide profiling of the study area was conducted by remote sensing of Quickbird (0.6m) satellite images and a 60-day field study to collect comparative data on: land use, vegetation type, damage to forest land, agricultural land, habitations, ownership, underlying causes of landslides, hazard assessment, and casualties. Data from Neelum valley's right-bank will be compared with data from its left-bank. This study was conducted by IUCN-P together with support from UNIL-IGAR.

4. A socio-economic survey of two villages to understand land use strategies and coping mechanisms of communities. Assisted by researchers from Islamabad-based Sustainable Development Policy Institute, an exploratory study was conducted in two villages using focus group discussions and semi-structured interviews. The discussions were facilitated by a close-up photo of the selected villages, (from the high resolution Quickbird image), providing a basis for discussing issues related to land use, risk and coping mechanisms.

Results

Information Needs

Meetings with a wide variety of stakeholders, decision-makers and communities underscored the interest and need for the GIAN-Pakistan study. Based on the interviews, we identified the following information needs and actions that would be useful for successful disaster risk reduction in EQ-affected AJK.

☆ Identification of the high risk areas;

☆ Determining which factors may aggravate future landslides: terraces, roads, grazing, damaged root systems, deforestation;

☆ Identifying possible mitigation actions;

☆ Assessing the role of vegetation in stabilizing slopes;

☆ Understanding techniques for landslide stabilization;

☆ Understanding community awareness of risk and coping mechanisms;

☆ How to institutionalize disaster risk management;

☆ How to integrate prevention measures in development projects and land use planning.

Landslide Susceptibility Model

Studying the link between deforestation and landslides area using spatial and statistical analysis, Pascal Peduzzi, Early Warning Unit, UNEP/GRID-Europe

Gathering evidence on parameters potentially connected with landslide occurrence over large remote areas can be time-consuming, hence the need to obtain a first assessment using global datasets and both GIS and remote sensing techniques. In particular, the goal of the exercise was to assess whether deforestation can lead to a higher occurrence of landslides.

To this end, a study was performed (Peduzzi, *in prep.*) using landslide areas (as detected by NESPAK and SERTIT by means of remote sensing) and available global datasets as well as low-cost DEM (from ASTER) and free Landsat TM 2000 imagery. GIS techniques were applied to extract contextual parameters (such as slopes, distance from fault lines, epicentres, rivers, roads and trails). Remote sensing was also applied to compute Normalised Difference Vegetation Index (NDVI). A statistical regression was then performed to identify what parameters best explain the size of landslides that were triggered by the 2005 Pakistan earthquakes.

Outputs include a map of landslide susceptibility (Figure 7.2) and a close-up providing visual evidence that forested areas are less susceptible to be affected by landslides than sparsely vegetated areas, other parameters being equal. The statistical results show that if slopes and proximity from the fault lines are the main contributors, followed by the presence of roads/trails, the study clearly revealed that area covered by forests suffered much fewer landslides than deforested areas.

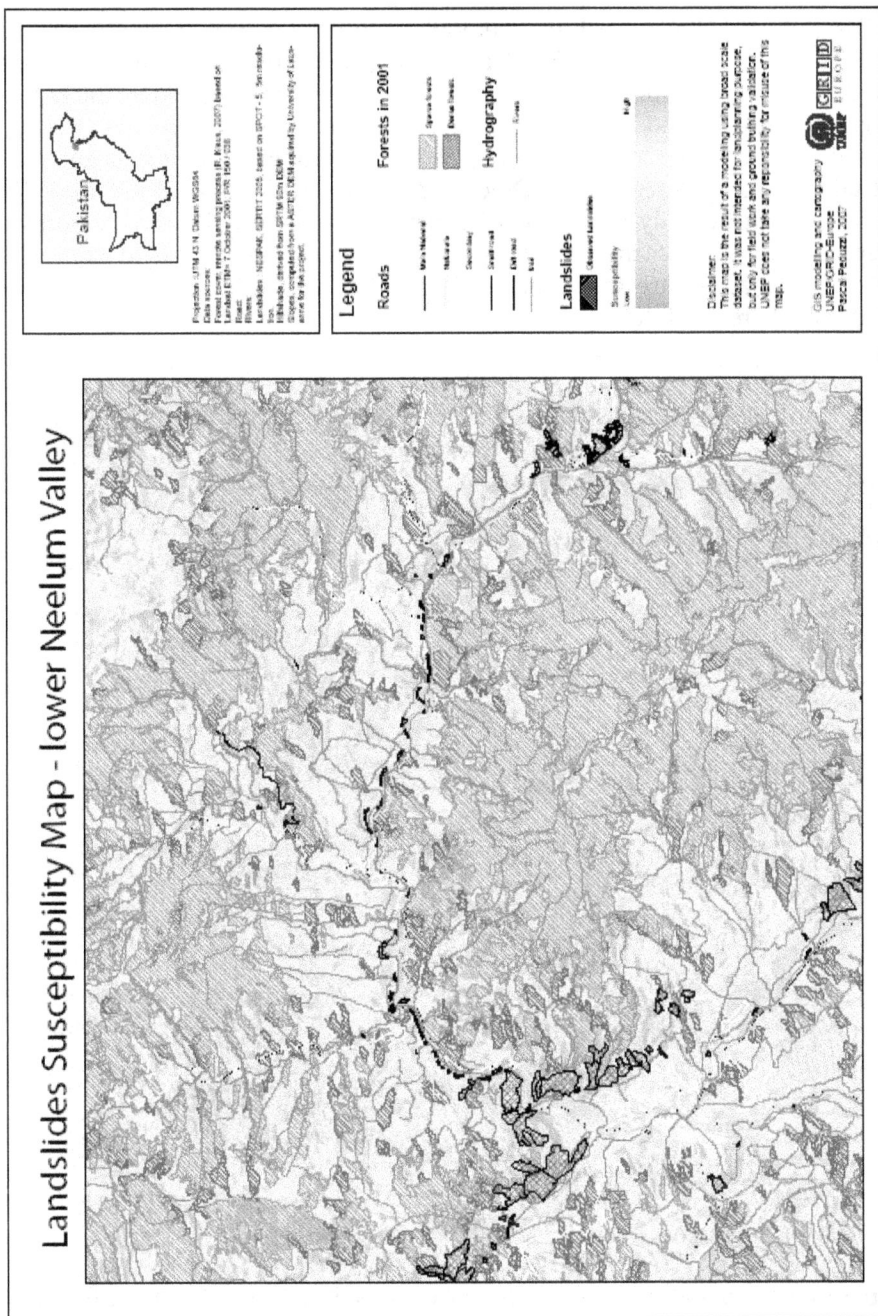

Figure 7.1: Landslide susceptibility shown in red, observed landslides shown in black and forest cover in green.

Hazard Profiles of the Study Area

Total Data Set: Landslides, Cracks Flood Zones

The study area, lower Neelum River Valley comprises approximately 381 m2 at 73°24'19 E, 34°30'7.2N and 73°36'43 E, 34°30'16N. Within this area, data were collected on 100 landslides and 17 crack zones, which were triggered by the EQ, as well as 3 flood zones and 24 landslides triggered by rainfall subsequent to the EQ. The small data set for flood zones will not be included for most of the data analysis. Landslides were analyzed on the Quickbird satellite image and cover all major landslides within the study area. The crack zones that were selected are not exhaustive, but represent those zones that were most obvious while conducting field work. The three flooding zones are those that caused the most damage to the northern Muzaffarabad area and that remain a great risk to the population.

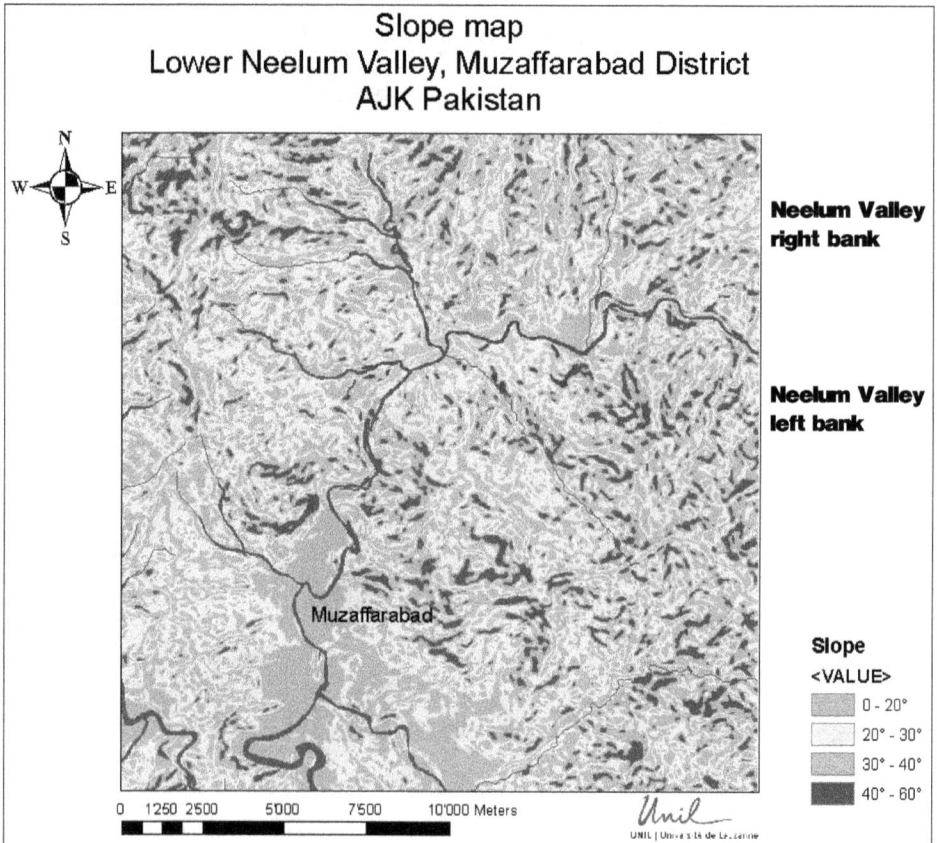

Figure 7.2: Slope map based on Digital Elevation Model (DEM).

Comparative Data Set between Neelum Valley's Left Bank (LB) and Right Bank (RB)

As outlined in the research design, the goal of the comparative data set is to compare landslide characteristics, land use, vegetative and damage assessments between lower Neelum Valley's RB and LB.

The data in Table 7.1 illustrate a geological difference between lower Neelum Valley's RB and LB and reflect the lower number of landslides, LB = 16, RB = 84. The RB is characterized by slump landslides, Murree formation and a lesser slope gradient than for the LB which experienced greater number of slump and fall slides of alluvium/colluvium/Abbottabad formation. See figure 2, Slope map for a visual representation of slope gradients. The surface area of landslides triggered by the EQ was significantly higher on the RB.

Table 7.1: Comparative data, geological characteristics.

	Landslides n=100	RB n=84	LB n=16
Type of landslide[1]			
Slump	62.0 per cent	69.0 per cent	25.0 per cent
Slump and fall	21.0 per cent	14.3 per cent	56.3 per cent
Fall	16.0 per cent	16.7 per cent	12.5 per cent
Slump and mudflow	1.0 per cent	0.0 per cent	6.3 per cent
Geology of landslides[2]			
Murree formation	57.0 per cent	60.7 per cent	37.5 per cent
Abbottabad formation	25.0 per cent	27.4 per cent	12.5 per cent
Lokahart/Hangoo/Kuldanna/Chorgali	9.0 per cent	8.3 per cent	12.5 per cent
Alluvium/Colluvium/mix with Abbottabad formation	9.0 per cent	3.6 per cent	37.5 per cent
Landslide slope average (degrees)			
Average	51	50	55
Min	25	25	35
Max	90	90	80
Surface area km²	17.02	13.45	3.57

1: The landslide categories are those used locally and are defined as follows:

Fall; free felled material

Slump; mass of land which slides in creeping movement

Fall and slump; started as fall but its lower section moved as slide

Mudflow; water saturated material

2: Murree formation : mix of sandstone, siltstone and shale (early Miocene era)

Abbottabad formation: dolomitic limestone (lower Cambrian era)

Kuldanna Formation (Greenish shales)

Lokahart Limestone, Hangoo formation, Chorgali Formation (Shales)

Table 7.2: Surface area of rainfall triggered landslides post EQ.

	Landslides n= 24	RB n = 22	LB n = 2
Surface area km²	0.427	0.335	0.092

Surface areas of rainfall triggered landslides were noted but other data were not collected due to time and budget restrictions. We note however, the similar pattern of higher frequency of rainfall induced landslides on the RB vs. LB.

Table 7.3: Comparative data of land use and ownership surrounding landslides.

	Landslides n=100	RB n=84	LB n=16
a. Landuse (multiple responses possible)			
Grazing/deforested	54.8 per cent	59.6 per cent	29.7 per cent
Terraces	24.0 per cent	23.5 per cent	26.6 per cent
Habitation	23.8 per cent	22.9 per cent	28.1 per cent
Forests	17.0 per cent	16.1 per cent	21.9 per cent
Water channel	14.0 per cent	15.5 per cent	6.3 per cent
Vehicle road	9.3 per cent	7.4 per cent	18.8 per cent
River	5.0 per cent	2.7 per cent	17.2 per cent
Reforested	1.5 per cent	1.8 per cent	0.0 per cent
Commercial	1.5 per cent	1.2 per cent	3.1 per cent
Footpath	1.0 per cent	0.9 per cent	1.6 per cent
Bridges	0.5 per cent	0.3 per cent	1.6 per cent
Landslides	0.5 per cent	0.6 per cent	0.0 per cent
Water supply	0.3 per cent	0.3 per cent	0.0 per cent
b. Ownership of land surrounding landslides			
Private	50.0 per cent	50.0 per cent	50.0 per cent
Private/*shamilat*	23.0 per cent	27.4 per cent	0.0 per cent
Private/government	17.0 per cent	14.3 per cent	31.3 per cent
Shamilat	0.0 per cent	0.0 per cent	0.0 per cent
Government	10.0 per cent	8.3 per cent	18.8 per cent

Data from Table 7.3 show a difference in vegetation cover and level of grazing surrounding the landslides on the LB compared to the RB, confirming the landslide susceptibility model and analysis of the Quickbird satellite image. A greater number of landslides on the LB seem to have been caused by vehicle roads as compared to the RB. Ownership differences confirm a difference in land tenure regimes, with a higher percentage of land under private/*shamilat* (state land under private management) on the RB as compared to the LB where the forests are more restricted to public use. However, the *type of state management* was not considered and could be the focus of further study. Another parameter that may be significant and which deserves further study is the role played by slope aspect and moisture gradient; as mentioned previously, the RB is a south-facing slope, it water retaining capacity may vary considerably with the LB, or north-facing slope.

Summary of findings

A majority of landslides (56 per cent) were caused by human-induced factors, especially deforestation and grazing, poor terracing and habitations located on exposed slopes and road construction, the remainder is related to proximity to rivers, steep slopes and geological features. The damage inferred by landslides in lower Neelum Valley amount to an estimated 72 million PKR ($USD 1.2 million), not including damage to the power supply, which amounted to 238 million PKR ($USD 3.9 million). The cost induced by the landslides constitutes a significant economic setback to the region and could have possibly been reduced by half with improved natural resources management.

Socio-economic survey

Land use strategies, risk perceptions and vulnerability: a qualitative survey of Saidpur and Kohori villages in Neelum Valley

Introduction

Social institutions, land tenure rights, access to resources, especially for women and economic considerations are key factors that determine how individuals manage their resources, or their land use strategies. Experience from other mountainous areas shows that a number of community land use strategies have evolved to reduce risk, such as spreading land holdings geographically or communicating warnings of imminent hazards, especially for flooding (Denkens 2007). In areas with rapid population increases and lack of suitable agricultural land, certain traditional risk reduction practices may no longer function. A successful mitigation programme may require changes in land use, *i.e.*, reforestation, retention walls, drainage schemes, improved terracing, improved management of communal lands etc. Such efforts can only be successful with the participation of communities (Cernea 1989). Local knowledge of risk, risk avoidance schemes and possible barriers to changes in land use strategies should be mapped before designing mitigation actions (Alacantara-Ayala 2004, Berkes 1999).

We need to clearly define the distinction between vulnerability and risk due to natural hazards. Poor communities face a number of factors that can make them vulnerable: crop failure, lack of fuel wood, lack of employment, droughts, access to clean water, disease, climate change and hazards such as landslides and flooding (Wisner *et al.*, 2003). Risk from a natural hazard exists when there is a threat to human lives, buildings or livelihoods. It should be seen as having two components, the likelihood of something adverse occurring and the consequences if it does (Crozier and Glade 2005).

Study Questions and Methodology

☆ Which land use strategies to reduce risk from landslides and flooding existed before the earthquake and how have these changed as a result of the earthquake?

☆ How do communities perceive risk due to natural hazards before the earthquake compared to post-earthquake?

The survey used focus-groups, semi-structured interviews (n=27), a discussion of a close-up map of the villages based on the Quickbird image and transect walks of the villages. Two female interviewers enabled discussions with females, which were equally represented in sample.

Saidpur village is approximately 17km northeast of Muzaffarabad, away from the main valley, and was considerably damaged from multiple landslides, both during the EQ and during subsequent rainfall. This village is more dependent on farming and grazing, combined with incomes from commuters to Muzaffarabad. Kohori village is the first village after Muzaffarabad and is more dependent on income from shopkeeping. It was more severely destroyed than Saidpur, due to the higher frequency of cinderblock constructions versus the traditional katcha houses in Saidpur and closer proximity to the epicenter.

Findings of the Socio-Economic Survey

The transect walks exposed a multitude of crack zones, which create a major risk factor during abundant rains that should be monitored, if possible by the communities. Many survey participants were aware of the need to drain water away from cracks and landslides however, we observed few examples of drainage. Risk perception of future landslides remains high in the villages surveyed, especially among women, compared to very little concern for landslides or earthquakes previously; previously flash flooding had been the main concern. Few respondents connected deforestation with the occurrence of landslides, rainfall being the most commonly mentioned cause. Communities have adapted to risk by abandoning exposed fields and houses and by reconstructing houses per ERRA standards (light materials yet poorly adapted to climate) but they may be forced to cultivate exposed fields as relief assistance dwindles. Most families are staying in villages, even if risks are high and usually due to lack of relocation options. The result is a reduction of arable land and a loss of income: men who migrated for work returned to villages to reconstruct their houses.

Observations and Recommendations

Improving Decision-Making Tools through an Interdisciplinary Approach to Understanding Landslides and Land Use

☆ Access to high-resolution satellite images and GIS software should be made free or at low cost not only in post-disaster situations but long-term for mountainous regions exposed to mountain hazards. Examples include the SERVIR Global Earth Observation System of System for Mesoamerica, a joint NASA/U.N. venture.

☆ Landslide susceptibility maps provide a larger scale view of areas susceptible to debris flows and the importance of land use variables, especially vegetation and road construction; training on how to produce such maps can be funded by donor and programmes aimed at capacity building for disaster risk reduction.

☆ Landslide susceptibility maps are based on models and are thus not appropriate for detailed planning for risk reduction. For this, detailed risk maps should be developed by knowledgeable agencies, with the proper authority to make policy changes, such as recommending the evacuation of high risk populations.

☆ Post-disaster assessment should include a simple methodology for gathering data on land use: grazing, terraces, deforestation, roads, habitations, ownership, economic damage, etc, in addition to typical geophysical assessments. This interdisciplinary approach to assessing landslides offers policy makers a more holistic picture of the land tenure situation, and underlying causes of landslides that should be the first step in designing a sustainable disaster risk reduction strategy.

☆ Conducting a post-disaster socio-economic survey to understand land use strategies, coping mechanisms, risk awareness, vulnerability issues is necessary in order to design community-level mitigation programmes and incorporate risk reduction into development activities.

☆ Special attention should be made to including women's perspectives in surveys on land use as women have a principal role in managing natural resources through firewood gathering and tending to livestock.

Natural Barriers and Risk Reduction in Mountainous Areas

☆ The role of protective forests, which are firmly established in some European countries, should be examined as cost effective natural barriers to disaster risk reduction in mountainous areas (Dorren *et al.*, 2007).

☆ The costs of maintaining forests for protection should be carefully weighed against the enormous cost of a full-blown debris flow or flooding event.

☆ Establishing community-level early warning systems and monitoring of cracks, landslides and flood areas are critical for reducing risk in mountains. These systems can be simple, for example working with the local religious leader to announce imminent threat via the town mosque, and establishing a stick and string method for checking on crack movement.

☆ Awareness about the dangers of cracks, need for drainage schemes and the link between vegetative cover and debris flow is a necessary component of disaster risk programmes, and should include private land owners, women's groups as well as managers of state-owned resources.

☆ Government policies to support the distribution and price of cooking gas as an alternative to firewood may have considerable positive influence on reducing deforestation rates.

☆ Awareness and training in locally adapted soil stabilization techniques, planting fast growing trees, shrubs and grasses on contours and slopes, proper terracing, retention walls and road construction methods can be effective disaster risk reduction components.

☆ Road construction is another major source of slope destabilization and any plans for new roads need to include proper grading and locally adapted techniques for slope stabilization such as placing vegetative mesh, combined with soil stabilizing plants. Exposed roads should receive priority attention from road authorities. Expertise on proper methods for soil stabilization can be obtained from IUCN- Pakistan.

☆ Other capacity-building measures for mountain communities should include community-level first aid training, emergency first aid kits, access to radios (if coverage is possible) blankets and food stocks, with special attention to women's needs.

☆ Integrating disaster risk reduction in programmes and institutions such as education, roads, health and natural resources management is necessary to improve disaster preparedness and recovery post disaster. This includes the importance of information and skills sharing among local partners.

☆ Enforcing zoning regulations to include risk from landslides, flooding and earthquakes in the construction of public buildings is a political issue but essential to disaster risk reduction.

☆ The evacuation of populations at high risk should be a given but is one where cultural and significant political issues come into play.

Conclusions and Future Research Needs

Natural disasters in mountains such as in Kashmir 2005 require an understanding of the underlying causes in order to design effective risk reduction programmes. For developing countries, the underlying causes of landslides are certainly linked to problems of economic development, poverty and resource

Figure 7.3: Locally adapted soil stabilization techniques.
Abdur Rauf Qureshi, IUCN-AJK 2007.

degradation. The goal of this study was to understand how to strengthen tools for decision-making for disaster risk reduction. Toward this goal, our objectives were to examine the links between land use and landslides using satellite images, landslide susceptibility modeling and on-site data collection, including a socio-economic survey of risk perception and land use strategies. This interdisciplinary approach to assessing landslides offers policy makers a more holistic picture of the underlying causes of landslides and an improved basis for designing a sustainable disaster risk reduction strategy.

Recommendations include the need to work with communities in establishing locally-adapted monitoring, mitigation and early warning systems; free or low-cost satellite images and GIS software, made readily available by donors and international organizations for all hazard-prone mountainous areas. GIS-based tools are essential for a spatial understanding of hazards; and data collection on landslides should go beyond geology to include land use, ownership and economic damage to provide a larger perspective on causes and mitigation options.

The study demonstrated a strong link between vegetative cover, ownership and forest management regime, terracing, road construction and debris flows. The policy implications are clear: a need to include improved resource management into risk reduction strategies, including building awareness and incentives for private owners to participate in increasing vegetative cover. Special attention should be given to the role of women as primary managers of natural resources. Road construction is another major source of slope destabilization and any plans for new roads need to include proper grading and locally adapted techniques for slope stabilization such as placing vegetative mesh, combined with soil stabilizing plants.

Of particular difficulty is the challenge of prevention. Unfortunately, history has proven that institutional and behavioral change is most likely to occur as a result of some type of shock, making prevention extremely difficult to implement. Although outside the scope of our study, some reflection should be given to the convergence between climate change adaptation and disaster risk reduction. Both signify human systems confronted with new environmental conditions, either through slow change, or quick on-set extreme events. More research on the protective qualities and thresholds to withstand extreme events is needed to convince policy makers about the critical link between disaster risk reduction and sound natural resources management. For example, the role of protective forests, which are firmly established in some European countries, should be examined as cost effective natural barriers to disaster risk reduction in mountainous areas, and deserves more research for implementation in a developing country context.

This is the challenge of governments, international organizations such as ISDR, UNEP, IUCN and donors – to act long-term and to push for prevention before disaster strikes. We hope that the study we have presented here adds to the growing literature on the need for preventive measures in mountainous regions: in particular improved natural resources management, adapted road building, monitoring and awareness building about mountain hazards. To this end, making tools such as satellite images, GIS software and training available to decision-makers and planners in disaster-prone regions may contribute significantly toward disaster risk reduction.

Acknowledgements

We would like to thank GIAN for funding this project, and for the encouragement and expertise of all partners who made this project possible.

References

Abramovitz, J., T. Banuri, P. Girot, B. Orlando, N. Schneider, E. Spanger-Siegfried, J. Switzer, A. Hammill (2002). *Adapting to Climate Change: Natural Resource Management and Vulnerability Reduction,* World Conservation Union (IUCN) Worldwatch Institute, International Institute for Sustainable Development (IISD) Stockholm Environment Institute/Boston (SEI-B).

Ahmed, J. and F. Mahmood (1998). *Changing Perspectives on Forest Policy, Policy that works for forests and people, Pakistan country study,* The World Conservation Union, Pakistan, Islamabad.

AJK Planning and Development Department, (2005). *Azad Kashmir at a Glance 2005.*

AJK Forest Department (2001). *Forestry Statistics of Azad Kashmir.*

Alacantara-Ayala, I. (2004). Flowing Mountains in Mexico, Incorporating local knowledge and initiatives to confront disaster and promote prevention, *Mountain Research and Development,* Vo. 24, No. 1 Feb. 2004 10-13.

Alexander, D. (2005). *Vulnerability to Landslides,* pp. 175-198 *In: Eds. Glade, T, M. Anderson, M. Crozier 2005. Landslide Hazard and Risk,* Wiley and Sons, Ltd.

Berkes, F. (1999). *Sacred Ecology. Traditional Ecological Knowledge and Resource Management.* Taylor and Francis, Philadelphia and London.

Bulmer, M. T. Farquhar, M. Roshan, S.S. Akhtar and S.K. Wahla, (2007). *Landslide Hazards After the 2005 Kashmir Earthquake,* EOS, American Geophysical Union, Vol. 88, No. 5, 30 January, 2007.

Cannon, T. (1994). "Vulnerability Analysis and the Explanation of "Natural" Disasters, In: Varley, A. (ed.) *Disasters, Development and Environment,* John Wiley and Sons.

Cernea, M. (1989). *User Groups as Producers in Participatory Afforestation Strategies,* World Bank Discussion papers 70, Washington, D.C.

Crozier, M. (1986). *Landslides: Causes, Consequences and Environment.* Croom Helm, London.

Crozier, M and T. Glade. (2005). *Landslide Hazard and Risk: Issues, Concept and Approach,* pp. 1- 40 In: Eds. Glade, T, M. Anderson and M. Crozier M. 2005 *Landslide Hazard and Risk,* John Wiley and Sons, Ltd.

Cruden, D. and D. Varnes,. (1996). Landslide Types and Processes pp. 36-71 *In Turner, K. Schuster, R (Eds) Landslides Investigation and Mitigation, Special Report 247,* Transportation Research Board, National Research Council, National Academy Press, Washington, D.C.

Denkens, J. (2007). Local Knowledge on Disaster Preparedness: A framework for analysis and data collection *Sustainable Mountain Development Vol. 52, Spring 2007.*

Dorren, L., Berger, F., Jonsson, M., Krautblatter, M., Mölk, M., Stoffel, M and A. Wehrli, 2007. "State of the art in rockfall – forest interactions." *Swiss Forestry Journal*, 6/2007.

Geological Survey of Pakistan (GSP) 2007a. *Study of geo-hazards triggered by 8th of October 2005 earthquake in Muzaffarabad, Jhelum, Neelum and Kaghan valleys*, GSP report, Islamabad, Pakistan.

Geological Survey of Pakistan (GSP) 2007b. personal communication.

Hewitt, K. (ed.) 1983 *Interpretations of Calamity from the Viewpoint of Human Ecology*, Allen and Unwin, Boston.

Hufschmidt, G. M. Crozier, and T. Glade, 2005. Evolution of natural risk: research framework and perspectives, *Natural Hazards and Earth System Sciences*, 5, 375-387.

International Strategy for Disaster Reduction (ISDR) 2004. *Living with Risk. A Global Review of Disaster Reduction Initiatives*. Geneva, Switzerland.

International Strategy for Disaster Reduction. (ISDR) 2007. *Environment and Vulnerability, Emerging Perspectives*, Environment and Disaster Working Group, Geneva Switzerland.

Klaus, R. (2007). *Hazards impacts and environmental changes: Research of potential links between deforestation and landslides in Northern Pakistan, An internship report*, UNEP/GRID-EUROPE.

Millennium Ecosystem Assessment (MEA), 2006. *Ecosystems and Well-Being: A framework for assessment*. www.millenniumassessment.org/en/Framework.aspx.

Peduzzi, P. (*in prep.*), *Modelling susceptibility of landslides occurrence and potential link with deforestation in North Pakistan*, Natural Hazards and Earth System Sciences.

Petley, D. Lin J-C. and JenC-H. 2006. On the Long-Term Impact of Earthquake-Triggered Landslides *International Conference on 8 October 2005 Earthquake inPakistan: Its Implications and Hazard Mitigation18-19 January 2006, Islamabad*.

Phillips, C. and M. Marden 2005. Reforestation Schemes to Manage Regional Landslide Risk, pp. 517-548 *In: Eds. Glade, T, M. Anderson, M. Crozier 2005. Landslide Hazard and Risk*, Wiley and Sons, Ltd.

Pryor, L. 1982. Ecological Mismanagement in Disasters, *The Environmentalist*, Vol. 2, Reprinted by: IUCN Commission on Ecology, the League of Red Cross Societies, IUCN, Gland, Switzerland.

Salvidar-Sali, A. and H. Einstein. 2007. "A Landslide Risk Rating System for Baguio, Philippines", *Engineering Geology* 91 (2007) 85-99.

Schneider, J. 2006. Earthquake Triggered Mass Movements in Northern Pakistan with Special Reference to the Hattian Slide *International Conference on 8 October 2005 Earthquake in Pakistan: Its Implications and Hazard Mitigation18-19 January 2006, Islamabad*.

Smith, K. 2004. *Environmental hazards: Assessing risk and reducing disaster*, Routledge, London, New York.

Sudmeier-Rieux, K. H. Masundire, A. Rizvi and S. Rietbergen (Eds.) 2006. *Ecosystems, Livelihoods and Disasters, An integrated approach to disaster risk management,* IUCN, Ecosystem management series, no. 4 Gland, Switzerland.

Wisner, B. P. Blaikie, T. Cannon and I. Davis. 2003. *At Risk, Natural hazards, people's vulnerability and disasters (Second Edition)* Routledge, London and New York.

Zingari, P.C. and G. Fiebiger, 2002. Mountain risks and hazards, *Unasylva,* No. 208, International Year of Mountains, FAO, Rome.

8

Using Empowerment Planning as a Strategy in Drought Management

Shweta Singh

Droughts

The problem of drought is defined as a "failure to receive expected precipitation for a period long enough to hurt" (Hare as cited by Dagel, 1997 p. 193). However, droughts have much more severe connotations than its definition leads us to believe. A large segment of the population feels the impact of drought in agrarian economies, such as India. The rain dependent nature of Indian agriculture and the large population that it supports, compound the problem. The victims of droughts are in a somewhat unique position as compared to victims of other natural disasters. Because the slow onset of drought conditions provides planners and affected groups with some time to anticipate the occurrence of a drought. However, it does not mitigate the pervasive and recurrent nature of drought (Gregory, 1989). Thus, the phenomenon of drought should ideally be a project for risk control through disaster management and planning. This paper presents an empowerment approach to planning for drought management. It presents a tool for collecting baseline data on droughts and an equation to calculate the risk (through hazard and vulnerability assessments), and coping competency at household, community, and regional level. It also presents an alternative set of indicators to measure potential outcome of drought related harm at these three levels.

Empowerment Approach in Disaster Management

Efforts to convince national governments to include a disaster component in policy planning have been largely unsuccessful. This is largely because planners have not been completely convinced of the effectiveness or cost efficiency of this proposition. This paper presents an empowerment based conceptual framework to facilitate program and policy planning for disasters as part of mainstream policy. Empowerment approach is drawn from a strengths based approach to working with marginalized clients in social work. Within the empowerment approach, client strengths are as important as their limitations. Additionally, the empowerment approach requires assessing the social ecology of the client problem at micro and macro levels and aims to strengthen client functioning at all levels. Using the empowerment approach this tool introduces coping competency in the equation for risk management. Furthermore, the tool proposes a region, community, and household level operationalization of risks and coping with drought.

Components of Empowerment Planning

☆ Empathy with (Potential) victims from Planners and Unaffected Community

☆ Managing risk – Skills and knowledge base

☆ Projects and policy additions and editions

☆ Women as the core instrument

☆ Emergency strategies

☆ Right focus - people or property or power

☆ Merging interests and resources

☆ National attention

Natural geography and repeated onslaughts of droughts make some areas more likely to be affected, such as Rajasthan. These areas are isolated from the rest of the country and are left to their own efforts in handling the problems. The country as a whole needs to empathize with the victims of recurring disaster like droughts. For instance, the suicides by the farmers in Andhra Pradesh, and the droughts that occur frequently in Orissa need to find a space in the collective conscience.

Secondly, after being repeatedly afflicted by droughts, the vulnerability of this population increases to make each occurrence in the future more challenging. The victims need to be provided with a knowledge and skill base that can be utilized towards coping with droughts. For instance, the affected population groups should be given a basic literacy on the use of scales and instruments for monitoring soil and climactic changes or cropping patterns to improve soil quality.

Thirdly, the current policy and programs need to be reexamined in light of what the data from previous droughts and its aftermath tells us. Additional policy and programs should be constituted that are contextualized to smaller groups of homogenous populations. For instance, policies like wasteland afforestation programs being undertaken with support from NABARD.

Fourthly, Gender issues suffer during any crises and women's weak position within households and community get adversely affected during droughts. Men's migration and dwindling resources put the burden of agriculture and child care solely on women (Agarwal, 2000). Gender specific planning in droughts can make women an asset to community coping skills. For instance, NGOs have been working with women in the Aravalli region of Rajasthan by giving them alternate employment in looking for water sources near village settlements.

Fifthly, drought should be treated as an emergency and adequate emergency measures, including resources, institutional collaboration, and rehabilitation services should be planned in detail.

Sixthly, the focus of intervention measures during drought should be clear at all levels of operation. The segregation in society and imbalance in power structures should be taken into account in order to ensure equity in services delivered. For instance, the drinking water scarcity continues in Gujarat while irrigation wells are exhausting under ground water tables (Mehta, 2001).

Seventhly, interstate collaborations should be encouraged, with creation of sister hospitals, sister schools, sister cities, that would be able to temporarily absorb the influx of displaced populations.

Finally, the drought should be treated as a national level issue in media and administration even when the effects are localized so as to have volunteers and participation from all over the country. This will ensure reduced conflict between social groups and exclusion of affected groups from mainstream society (Moench, 2002).

Mapping the Three Levels

A key aspect of this empowerment strategy is decentralization through mapping of risk and coping factors at different levels. Furthermore, strengths and limitations of dealing with disaster vary at different levels of region, community, and household. This assessment also needs to be done in a horizontal mapping of regions, communities, and households. Assets and liabilities need to be identified in the context of geographic and social location and extent of functioning. For instance, in coastal Andhra the key assets are land ownership, income generating equipment, and savings (Bosher, Penning-Rowsell, and Tapsell, 2007). While in Bijapur, in Karnataka, the primary assets are the openness of farmers to use of modern technology in farming (Vasavi, 1994).

The empowerment approach addresses issues of integration of efforts, sharing accountability, and conflict between the multiple stakeholders (Ingle and Halimi, 2007). The approach integrates regional, community, and household efforts and increases accountability by outlining the tasks and requirements at each level for controlling harm from drought. It also facilitates the interdependence of groups and institutions by creating a need for contribution from each level for effectiveness in mitigating and preventing drought. Thus, when drought occurs, the state will not be able to ignore the marginal households and community in favor of others in its rehabilitation efforts. Also, at the regional level, the NGOs and the community can hold regional authorities accountable to standards of quality by asking for carrying out of specific tasks. The regional authority on the other hand, can ensure that the allocation and control of resources is not significantly constrained by local social hierarchy at community level (Attwood, 2005).

The responsibility for planning and participation in drought management is politically invested at the region level, whereas community and household participation require social action and motivation. Household participation can be constrained by socio economic status, linkages with local NGO or other service provider agencies, identification with the interests of other households, and the belief that at least some members of their community and serving institutions are working for the household's benefit (Beard, 2007). Community participation can be constrained by the presence of heterogeneous composition, conflicting interests, and divisive dynamics within community (Adger, 2003). In view of this, community and household participation depends upon a realistic assessment of communities' initial skill and capabilities. Social cohesion in the community and equality within the household are also important to participation (Agrawal and Gibson, 1999). The localization of needs and role in assessment will also maintain the interest of community. Providing community and households with the right set of tools and information includes i) simple and relevant objectives of policy or program, ii) honest appraisal of policy or program in view of community feedback, and iii) facilitating the creation of specific tasks towards achieving program/policy goals.

Operationalizing the Three Levels

Region

Refers to the state level government and the institutions therein that have the larger responsibility and authority to determine policy and programs for development of the region and its people.

Community

Community refers to an interconnected collective of individuals and groups that identify similar needs and rely on a set of common resources, live in a geographical space that can be easily accessed by others in the collective, and have a combination of dynamic and static power differentials within the groups.

Household

Households refers to a group that lives in the same house, own common assets, labor, and have a share in earned income and its consumption, and are responsible for and accountable to each other in accordance with social norms and values that they jointly adhere to.

The Disaster Equations

Policy Hypothesis

Potential Outcome of Disaster = (Cost of relief and rehabilitation)

Empowerment Hypothesis

Risk (Hazard × Vulnerability) - Coping < Potential Outcome of Disaster/ (Social + Psychological + Economic + Political + Environmental costs)

The policy hypothesis centers on the costs to provide relief and rehabilitation to the affected group. This cost in the policy hypothesis is a function of services deemed essential and predetermined by policy makers at regional and national levels. The Empowerment hypothesis postulates that investment into reduction of risk, which is a function of hazard and vulnerability indices, and enhancement of coping capacity would be less than the outcome of disaster, when the potential harm from disaster is assessed using the cost of its social, economic, psychological, environmental, economic, and political manifestations.

Equation Components

Risk in the equation is a dynamic framework. It comprises hazard and vulnerability. Hazard is the likelihood of a disaster occurring . A "phenomenon that has the potential to injure life/livelihoods/and habitats" (AriyaBandhu and Wikramasinghe, p. 21. 2005). Vulnerability is the quotient for assessing the relative impact of the hazard if and when it occurs. Vulnerability is a "set of conditions that affect the ability of countries, communities, and individuals to prevent, mitigate, prepare for and respond to hazards" (AryaBandhu and Vikramsinghe, p. 23. 2005). Small hazard with a large vulnerability will compound the impact of the disaster. Coping is the capacity and competence to adapt to the change in the event of a hazard with minimal harm to wellbeing and productivity. It is important to distinguish between reducing vulnerabilities and enhancing coping, because they affect the dynamics of disaster differently and vulnerability and coping, each have a different set of indicators. Outcome of disaster is extensive including social, economic, political, psychological, and geographical and environmental outcomes. The indicators of drought in the tool contain a reference to these dimensions. The outcome is also dependent on frequency and length of disaster occurrence.

The above empowerment equation proposes that a proactive, decentralized planning strategy for droughts is more effective than the current practice of reactive disaster planning. The empowerment approach proposes disaster preparedness through a constant inflow of funds, local participation, and creation of a support infrastructure at three levels, to counter the effects of drought and threat of droughts. This extends the traditional reactive approach that is only focused on temporary rehabilitation.

From the point of view of effectiveness, functioning, and cost, the empowerment approach presented in this paper argues that management of disaster by improving coping competency, reducing vulnerability, and monitoring hazard is desirable. Handling disasters post fact is ineffective and dysfunctional as previous experiences have demonstrated. The typical response of "food, fodder, and employment" (p. 237, Phadke, 2002) are not adequate or lasting. For instance, NFCR, the National Fund for Calamity Relief is partially funded by the central government. The States can approach NCFR for relief and rehabilitation expenses. The complex procedure and the bureaucratic red tape affect the utilization of the fund money. Because of which, several State's have started teaming up to obtain additional funds that are not used for providing relief to drought victims.

Need for an Updated Tool

There is a need for a tool that strategies' drought in a framework of needs and action at three levels of region, community, and household for improving adaptability to droughts (Adger, 2003). The tool presented in this paper is a baseline assessment tool- for hazards, vulnerability and coping that contains indicators for different levels of drought occurrence. In addition, this tool can be used to identify gaps in current responses at policy and program levels in reference to droughts.

Table 8.1: Assessing hazard for drought

Region	Community	Household
Cropping pattern (unsustainable patterns; cash crops that drain soil)	Kind of occupation in agricultural	Reliance on natural resources for fuel
Climate (wet/dry)	Land use ratio	Cash crops only way for income enhancement
Soil type (retains water – coarse/fine textured) Soil status (low organic content)	Environment education	Below/ marginal poverty level existence
Relief (natural resources; natural water reservoirs; distribution of dry and wetlands; presence of ground water; Extent of dry areas; inaccessibility of parts)	Limited or no use of watershed farming	Chronic food insecurity
Vegetation (reducing cover)	Single source of water	
Annual Rainfall (timeliness, below above normal; Length of dry spells)		Access to drinking water
Land use ratio	Population	
Water quality	Absence of local point person for monitoring water levels; state of soil,	
Ground Water overexploitation	Equitable and diversified access and control over water resources across social groups and status	
Water Resource Management and encouragement of conservation practices	Absence of animal health centers	
Absence of local research institution for assessment of drought potential; water resources; and water resource management	Participation in monitoring climactic change	
Technology and environmentally sound agriculture experts in ministries and planning		
Absence of Active role for Forestry department		
Ongoing evaluation of historical, and comparative drought data		

Table 8.2: Assessing vulnerability quotient in drought.

Region Level	Community Level	Household Level
Lower levels of per capita income	Absence of local government institutions	Lack of Savings
Undiversified regional economy	Significant differences in economic status within community	Existing Debt
Significantly lower share in the national GDP	Limited political influence	Membership in the lower rung in the traditional caste hierarchy
Lower levels of literacy and education status for population	Single caste	Lack of Participation in political setup
Lack of Administrative flexibility	Absence of local (working) NGO networks	No of people in the household
Absence of proactive implementation of gender equity in programs	Strict adherence to gender rules and norms	Gender based discrimination affects savings, water use, and decision making process on farming issues, property, etc
		No of children
Lack of Administrative responsiveness	Absence of income generating groups	
		No of older adults
Lower level of literacy in women	Limited access to credit	Reliance on self employment
Age distribution in population and proportion of children and older adults (Under 10 and above 65 years)	Little control on state power structure	
		Level of social exclusion
Local community participation in defining and identifying drought conditions	Functioning participatory local governance	
Training of community in asset management and preservation	Access to financial market	
Absence of market recovery programs	Technological illiteracy of community / lack of familiarity with web resources such as the drought monitor	
Absence of representatives from commercial and regional banks in remote rural areas	Role of local, unmonitored money lenders	
Lack of gender based research on rural living and management	Extreme stratification between elements in population on basis of class, caste, and religion	

Contd...

Table 8.2–*Contd...*

Region Level	Community Level	Household Level
Encouraging commercial agriculture in water scarcity zones	Concentrated ownership of water resources	
Lack of recognition of rural realities (and sub groups of farming, tree growing, and animal husbandry clientele) in universal commercial credit policies of rural banks, such as NABARD	Absence of ecological and conservation farming techniques.	
Lack of institutional cooperation across regions, such as limited data exchange		
Water conflicts with neighboring regions		

Table 8.3: Assessing coping quotient in drought.

Region Level	Community Level	Household Level
Interstate crises collaboration agreement	Trained and functioning Ngo with access to large emergency funds	Diversified income sources
Ratio of Disposable budgetary allocation for emergency and historical data on affected people and production	Community Center for counseling and food, and shelter	Belief in accountability of government departments
Trained taskforce of government officers, community volunteers and NGOs with clear division of micro task management	Emergency cross community networks (across caste/ religion, and class)	Informal support systems spread outside the region (though marriage and trade, culture)
Risk Mapping		Women's participation in income generating programs that do not rely on agriculture
Promotion of Policy to provide risk insurance and coverage for agriculture	Local level access to risk insurance	Use of insurance
Simple and mandatory coverage for household risk insurance	Credible and Responsible Ngos involvement in assessing and creating financial credibility of potential clients for control over insurance, for access to land, for credit from state institutions, and local control over local resources such as village common lands	Financial education on assets and income maintenance

Contd...

Table 8.2–*Contd...*

Region Level	Community Level	Household Level
Presence of shelters and need based supervision by Para military forces		Financial information flow about schemes for credit and agriculture support
Proofed plan of provision of food to inner parts of region	Tested plan for emergency	
Monitoring and maintaining wage rates for small farmers to encourage savings	Emergency task force	
Create temporary pockets with private & public collaboration for active wage monitored labor markets to absorb the temporary immigration	Active women participation on task force	
Development of alternative geographical locations through wasteland development for temporary or permanent and planned resettlement of affected population groups	Group cohesiveness, informed and responsible leadership to coordinate community responses to state and NGO initiatives	Access to wellness and counseling centers within community
Develop collaborations with social institutions, such as Schools and Hospitals to handle the migrant/ affected community for both outreach and in-house institution services		

Table 8.4: Indicators for Drought Outcome Calculation

Region Level	Community Level	Household Level
Loss of production	Displacement of community	Loss of life
Falling per capita income	Breakdown of social structures such as households and families	Loss of income
Loss of investment	Depleted resource base	Depletion of savings with 'consumption smoothing'
Rising food prices and inflation		Reducing consumption of essential nutrients
Political and social Instability	Loss of social capital through breakdown of social connections and networks	Likelihood of falling into the poverty trap
Rehabilitation costs over long period of time of displaced population	Dysfunctional administrative structures	Loss of assets
Stress on urban resources	Distrust within community	Loss of production goods

Contd...

Table 8.2–*Contd...*

Region Level	Community Level	Household Level
Increase in urban crime	Increased susceptibility to exploitation of community by urban/ neighboring region/ outsiders, such as through prostitution of women and children	Loss of livelihood
Potential conflict/ riots between urban and rural dwellers		Gender issues surface and women in worse off position with care of children and broken down farms and out migration of men
Distrust towards state and its institution		Psychological dysfunction
Import of essential food grains		Costs to health of farming, & dairy animals
Falling foreign exchange ratio and international debt		Deterioration of physical health of all household members and consequently wellbeing and productivity over the life course
Instability in internal financial markets		
Abandoned and semi abandoned rural areas serve as breeding and settling ground for individuals and groups that are a potential threat to state and civil society		

Using the Tool

For Research

The research agenda drawn from this tool would include baseline and longitudinal studies, mixed-method studies of relationship between the different sets of indicators, and predictive models that would be able to ascertain investment priorities for increasing coping and reducing risk. Research should also monitor the failure of some communities and households to collect baseline information. This process by itself will support the identification of an additional vulnerability – that of inability to participate in collective action/or absence of social capital (Beard, 2007).

For Programs

This tool should be used to identify groups on the basis of common vulnerabilities and coping skills in reference to droughts. This tool can also assist in determining cohesiveness between programmatic components at household, community and regional levels. This tool can be used as an assessment tool for the economics of disaster planning.

In rural areas where the definition of poverty still encompasses basic need of food, shelter (pucca/non leaky roof), limited debt, and few clothes (Krishna, 2006), empowerment can be the only approach to a disaster like drought. The strain of a drought at this level of existence and limited access to resources pulls households and entire communities into extreme poverty (Krishna, 2006). The focus of planning and the operationalization of its process need to be targeted to at least the three levels presented here, household, community, and the region. Additionally, drought management strategy itself needs to be broken down into outcome based monitoring of hazard, reduction of vulnerability, and enhancement of coping capability at the levels.

References

Adger, W. N. (2003). Social capital, collective action, and adaptation to climate change. *Economic Geography, 79*(4), 387-404.

Agarwal, B. (2000). Conceptualising environmental collective action: why gender matters. *Cambridge Journal of Economics, 24*(3), 283-310.

Agrawal, A., and Gibson, C. C. (1999). Enchantment and disenchantment: The role of community in natural resource conservation. *World Development, 27*(4), 629-649.

Ariyabandhu, M. M., and Wickramasinghe, M. (2005). Gender dimensions in disaster management: A guide for South Asia. New Delhi: Zubaan.

Attwood, D. W. (2005). Big is ugly? How large-scale institutions prevent famines in western India. *World Development, 33*(12), 2067-2083.

Bosher, L., Penning-Rowsell, E., and Tapsell, S. (2007). Resource accessibility and vulnerability in Andhra Pradesh: Caste and Non-Caste Influences. *Development and Change* 38(4), 615–640.

Beard, V. A. (2007). Household contributions to community development in Indonesia. *World Development, 35*(4), 607-625.

Dagel, K. C. (1997). Defining drought in marginal areas: The role of perception, *Professional Geographer, 49*(2), 192.

Gregory, S. (1989). The changing frequency of drought in India, 1871-1985, The Geographical Journal, 155 (3), 322-334.

Ingle, M., and Halimi, S. (2007). Community-based environmental management in Vietnam: The challenge of sharing power in a politically guided society. *Public Administration and Development, 27*(2), 95-109.

Kinsey, B., Burger, K., and Gunning, J. W. (1998). Coping with drought in Zimbabwe: Survey evidence on responses of rural households to risk. *World Development, 26*(1), 89-110.

Krishna, A. (2006). Pathways out of and into poverty in 36 villages of Andhra Pradesh, India. *World Development, 34*(2), 271-288.

Mehta, L. (2001). The Manufacture of Popular Perceptions of Scarcity: Dams and Water-Related Narratives in Gujarat, India. *World Development, 29*(12), 2025-2041.

Moench, M. (2002). Water and the potential for social instability: Livelihoods, migration and the building of society. *Natural Resources Forum, 26*(3), 195-204.

Parnwell, M. J. G. (2006). Eco-localism and the shaping of sustainable social and natural environments in North-East Thailand. *Land Degradation and Development, 17*(2), 183-195.

Vasavi, A. R. (1994). Hybrid times, hybrid people: Culture and agriculture in South India. *Man, 29* (2). 283-300.

9

Application of Locally Available Mulches for Moisture Conservation in Afforestation Programme of Dry Areas

Kumud Dubey, Anubha Srivastav and V.K. Singh

Introduction

The forest of Shankargarh range of Allahabad Forest Division is mixed dry deciduous type and extensive open cast silica mining activities have resulted into long barren unproductive lands causing great damage to the forest as well as productivity of the region. Since the selected site has low moisture content and high temperature during most of the period of the year. In the present study different locally available mulches were tried in afforestation programme of this degraded forest area. The Application of mulch improves the moisture retention capacity of the soil thus increase the water regime available for the plant.

Keywords: *Silica mining, Mulching, Reclamation, Moisture conservation, Afforestation, Water conservation.*

Mining is essentially a destructive developmental activity, where ecology suffers at the alter of economy. Unfortunately, in most regions of the earth, the underground geological resources are superimposed by biological resources *i.e.* forests. This is particularly evident in India. Hence mining operations necessarily involve deforestation, habitat destruction, biodiversity erosion and destruction of geological records, which contain information about past biodiversity. Extraction and the processing of ores and minerals also lead to widespread environmental pollution.

However, mankind cannot afford to give up the underground geological resources that are the basic raw materials for development. An unspoiled nature can provide ecological security to people but cannot bring economic prosperity. India is among the top ten mineral producing nations in the world and its economy depends on the value of mineral produced. Small-scale mining is more prevalent in India. Scientific mining operations accompanied by ecological restoration and reclamation of mined wastelands and judicious use of geological resources, with search for eco-friendly substitutes and alternatives must provide the answer. From the literature it follows that serious and sincere reclamation measures based on scientific research findings have not been taken in most of the mined areas. Moreover this revegetation of the mining areas of the country is not achieving desired success because of high rate of mortality and slow growth of planted species. The major problem in the reclamation of these mined lands is the initial establishment of the seedlings owing to sever moisture stress, specially during the post mansoon period. The present study is on the application of various mulches in the reclamation of mining lands.

An important mine of Silica, a major mineral used in glass industry, is situated in Vindhyan Hills of Uttar Pradesh state and its extensive quarrying and open cast mining have resulted into long barren, unproductive and deeply irregular sloppy lands. The soil has low moisture retaining capacity and the plantation in the reclamation programme suffers a moisture stress condition. To minimize this unfavorable situation, locally available mulches were used in the reclamation programme.

Materials and Methods

Site selected for the study was Shankargarh in Allahabad District. The area was degraded due to open cast silica mining and other biotic interferences. Area involved in mining in Shankargarh includes 40 villages covering about 150 sq. km. Over burden in the area is not uniform and varies from place to place. Since the selected site has low moisture content and high temperature during most of the period of the year. To conserve moisture of the soil, four treatments of mulches were tried for studying their effect on growth performance and survival of planted species. The mulch treatments were: mulch of wheat straw, mulch of rice husk, mulch of dry leaves of *Butea monosperma* and control (without mulch). These mulches are locally available. The species planted at reclamation site were *Pongamia pinnata, Prosopis juliflora, Acacia nilotica, Albizia procera, Azadirachta indica, Madhuca indica* and *Pithecellobium dulce*. These species were selected on the basis of choice of the local people as studied during socio economic survey and ecological condition of the site.

The plantation trial was established in Split Plot (factorial) Design under rain fed conditions with spacing of 2m x 2m. The pit size for plantation was 60 cm x 60 cm x 60 cm. The dug up soil of the pit was mixed well with 1 kg FYM and NPK mixture in ratio of 1:1:1. The seedlings were irrigated immediately after planting. The mulches of wheat straw was procured from the local villagers. Plenty of rice husk was available in the vicinity of rice mills. Since it does not have any other utility of economic importance at time it poses serious problem for its disposal due to its bulkiness. *Butea monosperma* trees are abundantly present in the area; collected leaves were dried and coarsely powdered for the purpose. The mulches used were applied in the quantity of 2 kg per plant every year after rainy season. The growth attribute *viz.* height, girth and survival percentage were recorded annually.

Results and Discussion

The growth data *viz.* height, girth and survival percentage of the plantation trial for studying the effect of different mulches on growth performance and survival of selected species under rain fed condition. are given in Tables 9.1 and 9.2 respectively. Results were statistically analyzed and found significant. However the effect of mulch treatments on survival of *Prosopis juliflora and Pithecellobium dulce* were not statistically significant, but effect of mulching on growth (height and collar circumference) is significant. The result (Table 9.1) shows that growth performance *viz.* height and collar circumference in *Pongamia pinnata* plants were found superior in wheat straw treatment followed by other treatments as compared to control. Similarly, in *Prosopis juliflora, Albizia lebbek, Acacia nilotica* and *Azadirachta indica,* growth performance was found best with wheat straw treatment as compared to control followed by other mulches. In *Pithecellobium dulce* and *Madhuca indica,* mulch of Butea leaves has shown better results over other treatments. Out of the seven species planted the specie performed well were: *Prosopis juliflora, Pithecellobium dulce, Azadirachta indica* and *Pongamia pinnata.* . From the result it was observed that mulching caused considerable improvement in tree growth. These practices caused better growth and productivity. The mulch of wheat straw was found most suitable for the purpose.

The challenges caused by an ever increasing population growth, coupled with an emerging water crisis in many parts of the world, require a new approach in managing land and water resources. This, water crisis in many parts of the world, threats to peace and security due to disputes on the sharing of water resources. India is also facing the same problem. This shortage of water in India is mainly due to explosive population growth, global warming, deforestation and depletion of water in underground storages (aquifers).To tackle this problem, issues of land use planning such as deforestation, soil erosion and water shed management should receive attention. Increasing the forest/vegetation coverage is an important approach for solving such problems. To improve the vegetation cover and to control the soil erosion in the degraded reserved forest and the abutting community lands, several afforestation programme are being run by the government. But these afforestation programme are not achieving the desired success because of the high rate of mortality and slow growth of planted species. The major problem in these vegetation programmes is the initial establishment of the seedlings due to sever moisture stress, especially during the post mansoon period. In the present study mulching is used to conserve the moisture in the afforestation programme of the dry areas. The use of crop residue mulch is an important strategy to reclaim a degraded land land (Panwar *et al., 2000;*

Table 9.1: The effect of different mulches on growth performance of planted species in silica mined area.

Species	Mulch Treat.	Initial Growth		After One Year		After Two Year		After Three Year		After Four Year		Growth Increment after Four Years	
		Av. Ht. (cm)	Av. cc. (cm)	Av. Ht. (cm)	Av. cc. (cm)	Av. Ht. (cm)	Av. cc. (cm)	Av. Ht. (cm.)	Av. cc. (cm)	Av. Ht. (cm)	Av. cc. (cm)	Av. Ht. (cm.)	Av. cc. (cm.)
Pongamia pinnata	M$_0$	28.20	3.00	78.50	4.70	94.00	4.02	142.37	6.00	187.50	9.50	159.30	6.50
	M$_1$	28.80	3.60	113.00	7.50	150.50	9.19	210.50	14.40	318.80	19.60	290.00	16.00
	M$_2$	26.50	3.00	80.90	5.50	139.00	7.66	178.00	9.20	270.19	13.50	243.69	10.50
	M3	28.20	3.10	74.60	5.30	103.00	5.25	161.00	8.34	230.00	12.40	201.80	9.30
Prosopis juliflora	M$_0$	104.00	5.00	268.00	10.50	337.00	12.86	420.00	15.00	475.00	17.00	371.00	12.00
	M$_1$	112.00	5.20	300.00	11.20	410.00	15.37	535.25	18.78	618.00	23.40	506.00	18.20
	M$_2$	99.20	4.70	261.00	9.80	353.00	13.41	478.40	15.88	542.00	18.50	442.80	13.80
	M$_3$	116.00	5.20	283.00	11.50	375.00	15.04	496.00	17.98	565.00	20.24	449.00	15.04
Acacia nilotica	M$_0$	70.20	3.20	82.80	5.10	73.00	5.14	95.10	6.50	112.00	7.60	41.80	4.40
	M$_1$	72.20	3.20	86.30	5.80	96.00	5.66	140.56	7.85	173.50	9.50	102.30	6.30
	M$_2$	80.40	3.80	95.90	6.90	91.00	7.14	122.00	9.30	138.89	12.89	58.49	9.09
	M3	72.70	3.50	92.30	6.50	100.00	6.61	155.45	8.60	170.80	11.90	98.10	8.40
Albizia procera	M$_0$	50.80	2.70	56.70	4.40	62.00	4.58	110.00	6.15	148.20	8.00	97.40	5.30
	M$_1$	36.50	2.50	51.90	4.10	74.00	4.96	165.00	6.98	240.30	8.50	203.80	6.00
	M$_2$	28.10	2.00	34.80	2.70	46.00	5.81	101.00	8.00	145.00	9.24	116.90	7.24
	M$_3$	32.20	2.40	54.60	3.80	48.00	6.70	105.00	9.52	152.00	10.45	119.80	8.05
Azadirachta indica	M$_0$	32.00	2.10	31.40	1.90	54.80	4.00	82.90	9.56	113.00	6.90	81.00	4.80
	M$_1$	38.10	3.00	55.10	3.80	75.00	5.16	125.70	6.80	182.00	8.61	143.90	5.61
	M$_2$	40.40	3.10	58.30	4.30	74.90	6.06	118.60	7.65	168.50	8.50	128.10	5.40
	M$_3$	32.60	3.20	58.60	4.30	71.00	5.34	111.92	7.01	154.40	8.20	121.80	5.00

Contd...

Table 9.1–*Contd...*

Species	Mulch Treat.	Initial Growth		After One Year		After Two Year		After Three Year		After Four Year		Growth Increment after Four Years	
		Av. Ht. (cm)	Av. cc. (cm)	Av. Ht. (cm)	Av. cc. (cm)	Av. Ht. (cm)	Av. cc. (cm.)	Av. Ht. (cm.)	Av. cc. (cm)	Av. Ht. (cm)	Av. cc. (cm)	Av. Ht. (cm.)	Av. cc. (cm.)
Madhuca indica	M_0	26.80	2.60	29.50	3.30	36.00	2.80	50.20	3.50	67.90	4.45	41.10	1.85
	M_1	18.20	2.50	23.50	2.70	32.00	2.90	47.50	3.89	81.50	4.95	63.30	2.45
	M2	18.00	2.50	25.70	2.90	31.60	2.79	45.80	3.40	65.70	4.80	47.70	2.30
	M3	12.60	1.00	16.10	1.90	38.00	3.18	57.31	4.20	84.50	5.25	71.90	4.25
Pithecellobium dulce	M_0	62.30	2.80	105.50	5.60	166.00	7.92	230.00	8.75	278.00	10.80	215.70	8.00
	M_1	62.60	3.10	109.10	5.80	191.00	9.40	295.00	11.50	345.00	13.79	282.40	10.69
	M_2	56.00	3.00	116.10	5.80	217.00	10.55	350.00	13.80	397.00	15.90	341.00	12.90
	M_3	57.00	2.50	102.00	5.10	192.00	11.61	310.00	15.00	340.00	18.50	283.00	16.00

Ht.: Height; Cc.: Collar circumference; M_0: Control; M_1: Mulch of wheat straw; M_2: Mulch of rice husk; M_3: Mulch of dry leaves of *B. monosperma*.

Table 9.2: The effect of different mulches on survival percentage of planted species after four year of plantation in silica mined area under rain fed condition.

Sl.No.	Species	Mulch Treatments			
		Control	WS	RH	BL
1.	Pongamia pinnata	16	50	38	40
2.	Prosopis juliflora	90	98	94	96
3.	Acacia nilotica	22	44	66	70
4.	Albizia procera	12	42	32	30
5.	Azadirachta indica	26	76	70	60
6.	Madhuca indica	18	30	20	26
7.	Pithecellobium dulce	82	88	84	84

WS: Mulch of wheat straw; RH: Mulch of rice husk; BL: Mulch of dry leaves of *Butea monosperma*.

Singh and Singh, 1999; Mishra *et al.*, 1996; Munir *et al.*, 1998 a, b.). It reduces evaporation, conserve moisture, controls weeds and reduces excessive heating(Bhattacharya and Mitra, 2000; Budelman-A.,1989; Rahman and Khan, 2001; Patil-SN *et al.*, 1991; Khan-MA,1989; Sharma *et al.*, 2001; Kumar Dinesh *et al.*, 2003). Its application may also increase soil health both through stimulation of micro biota beneficial to plant nutrient uptake and through addition of organic matter to soil after its decomposition. Mulching caused considerable improvement in plant growth and in turn the productivity (Sharma-PK *et al.*, 1990; Sharma-NK *et al.*, 1998; Uniyal-SP,*et al.*, 1994). These practices caused better growth and proliferation of root system which is essential for speedy establishment of plantation and imparting tolerance against droughts and famines. It may be concluded that the moisture holding capacity of the soil (Tomar-VPS *et al.*, 1992) was increased after application of various mulches, thus, better growth results could be achieved in such dry areas.

References

Bhattacharya, B.K. and Mitra, S. (2000). Moisture conservation in situe using biomulches in rain fed upland late sown Kharif moong. *Environment and Ecology*, 18(2): 375-379.

Budelman, A. (1989). The performance of selected leaf mulches in temperature reduction and moisture conservation in the upper soil stratum. *Agroforestry-Systems.* 8(1): 53-66.

Khan, M.A. (1989). Influence of tillage methods and mulches on soil moisture and yield of gram and wheat under rainfed condition. *Annals of Arid Zone*, 28(3-4): 277-283.

Kumar, Dinesh, Singh, Ranjeet, Gadekar, H., Patnaik, U.S., Kumar, D. and Singh, R. (2003). Effect of different mulches on moisture conservation and productivity of rainfed turmeric. *Indian Journal of Soil Consevation*, 31: 141-44.

Mishra, A.K., Bhowmik, A.K. and Banerjee, S.K. (1996). Effect of mulches on growth of tree species on fly ash. *Environment and Ecology*, 14(2): 411-414.

Munir, A.D., Majid, N.M., Abdol, I. and Khan, G.S. (1998). Effects of mulching on the growth of interplanted *Acacia mangium* on sandy tin-tailings in Peninsular Malaysia. Lyallpur Akhbar, 65: 35.

Munir, A.D., Majid, N.M., Abdol, I. and Khan, G.S .(1998). Effects of *Acacia mangium* pruning mulch on the growth of *Arachis hypogaea* and *Setaria splendida* intercrops on sandy tin-tailings in Malaysia. Lyallpur Akhbar, 66: 3.

Panwar, Pankaj, Bharadwaj, S.D. and Panwar, P. (2000). Effect of soil amendment and mulch on establishment of trees in sandstone mine spoils in mid hill zone of Himachal Pradesh. *Annals of Forestry*, 8(2): 214-219.

Patil, S.N., Morey, D.K. and Pore, D.B. (1991). Effect of vertical mulching on moisture conservation and yield of sorghum-pigeonpea intercropping. *Annals of Plant Physiology*, 5(1): 76-80

Rahman, M.S. and Khan, M.A.H., 2001. Mulching induced alteration of microclimatic parameters on the morpho-physiological attributes in onion (*Alium cepa* L.). *Plant Production Science*, 4(3): 241-248.

Sharma, P.K., Kharwara, P.C. and Tewatia, R.K. (1990). Residual soil moisture and wheat yield in relation to mulching and tillage during preceding rainfed crop. *Soil and Tillage Research*, 15(3): 279-284.

Sharma, N.K., Singh, P.N., Tyagi, P.C. and Mohan, S.C. (1998). Effect of *Leucaena mulch* on soil-water use and wheat yield. *Agricultural Water Management*, 35(3): 191-200.

Sharma, N.K., Singh, P.N. and Tyagi, P.C., (2001). Effect of application of Leucaena mulch on soil moisture cosevation and productivity of rainfed wheat. *Indian Journal of Soil Conservation*, 29(2): 143-147.

Singh, A.K. and Singh, R.B. (1999). Effect of mulches on nutrient uptake of *Albizia procera* and subsequent nutrent enrichment of coal mine overburden. *Journal of Tropical Forest Science*, 11(2): 345-355.

Tomar, V.P.S., Narain, P. and Dadhwal, K.S. (1992). Effect of perennial mulches on moisture conservation and soil-building properties through agroforestry. *Agroforestry Systems*, 19(3): 241-252

Uniyal, S.P., Tripathi, N.C., Singh, R.V., Shekhawat, G.S. (ed.), Khurana, S.M.P. (ed.), Pandey, S.K. (ed.) and Chandla, V.K. (1994). Effect of different locally available mulches on moisture conservation, yield of off-season potato under mid hills. Potato: present and future. Proceedings of the National Symposium held at Modipuram during 1-3, March, 1993, pp. 147-150.

10

Conservation and Utilization of Rainfed areas through Agroforestry interventions

Ram Newaj and Shabir Ahmad Dar

Introduction

Agroforestry has a great potential to utilize in rainfed areas and those are fallow, marginal, sub marginal and wastelands. The gainful utilization of off-season natural resources including precipitation and solar radiation takes place in agroforestry. In the long run agroforestry impart stability to rainfed agriculture and improve farmer's income apart from yielding a variety of products. Perennials (tree/shrub) are important components of agroforestry and in this system, medicinal, aromatics and dye yielding plants that are annuals can also be included in the production system, which having important role in human health and nutrition. Agrisilviculture (intercropping with NFTs, and alley cropping), dryland horticulture, hortipasture, silvipasture and tree farming/block plantations are the important systems by which the rainfed areas can be utilized efficiently.

Agroforestry is one of the most important biological methods for conservation of natural resources (soil, water and vegetation). In this system, tree and crop/grasses provides a good soil cover which protect the soil from run-off losses, add organic matter in soil by which infiltration rate of soil improve, besides soil fertility improvement. They also provide opportunities for conserving water by increasing infiltration and reducing transpiration. The integration of animals in agroforestry adds a further element to the nutrient cycle through addition of dung and urine for improving the soil nutrient status besides improving its structure, infiltration and permeability. In fact the potential for nutrient recycling through trees is reduced to a great extent because of use of such materials for fodder. However, animal integration compensates it through farmyard manure. The capacity of trees to grow under difficult climatic and soil conditions and their potential for soil and water conservation gives agroforestry an edge for practicing it on marginal lands, semiarid and arid lands, sloping lands and problem soils. Also, there is a demonstrated potential for reclamation of degraded lands. The evidence about ameliorative effects of trees on soils comes from comparing the soils beneath tree canopies and within the orbit of their root systems with the soils in the surrounding area beyond the influence of the trees.

India having 97 m ha rainfed area out of total 142m ha of arable land and it contributes 44 per cent food and supports 40 per cent of the population. Most of the coarse cereals except maize are grown in rainfed regions. These regions support 84 per cent groundnut, 91 per cent pulses, 55 per cent rice, 68 per cent cotton and 22 per cent wheat. Productivity of rainfed crops continues to be low; it is one half to one third of that obtained with irrigated crops. Low rainfall or failure of monsoon rains is a recurring feature of rainfed areas. This has been responsible for drought and famines. Prolonged deficiency of soil moisture adversely affected crop growth indicating incidence of agriculture drought. It is the result of imbalance between soil moisture and evapo-transpiration needs of an area over a fairly long period as to cause damage to standing crops and to reduce the yields.

Under such situations, alternate land use systems are able to make effective use of pre-or-post-monsoon rains due to their woody perennial components (trees and shrubs); because woody perennials having deep root system and they can efficiently utilize 30 to 40 per cent of the rainwater that goes deep into the soil where soils are predominantly light textured. Further, tree root systems help increase water infiltration into the soil and also intercept, absorb and recycle nutrients in the soil solution which otherwise have been lost in leaching, thereby making the nutrient cycle more closed. Another advantage of the perennial components in the system are; it reduces the risk of crop failure because tree can yields some things like fruits, fuel and fodder under adverse weather conditions when seasonal crop totally fail.

In order to have a sustainable production system, it is necessary to conserve and use properly our natural resources on which production is depending. Ways and means have to be found to achieve the above objectives through an agroforestry, which is more conservation-effective than an arable land use system. Although experimental evidences to support the logic are sparse, an attempt has been made to highlight results of few studies on agroforestry systems which imply continuous presence of both annual and perennial groups of plants simultaneously on same site, like agrisilviculture, agrihorticulture, agri-silvi-horticulture, hortipasture and silvipastoral system.

Agroforestry Potentials in Soil and Water Conservation

Water is a precious input in crop production especially under rainfed situations. Water erosion causes a severe loss of productive surface soil and eroding several million tonnes of soil annually. In order to have a sustainable production system it is necessary to minimize runoff and evaporation losses and increase infiltration of water into the soil. There are several options available for conservation of water resources-physical/structural and biological being the main contenders. Lot of research and development work has hither to been done on conservation of soil and water through physical structures. Varied successes have been achieved on this score. They are, however, costly and not long lasting, requiring higher skills for designing and development. It has been realized, of late, that more effective and sustainable conservation both of soil and water could be achieved through biological methods. The role of trees, shrubs and grasses in reducing/controlling erosion is well recognized. Trees act as barriers or as vegetative cover, thereby reducing/controlling erosion. The barrier function is operative by reducing wind speed, checking runoff and suspended sediments. The cover function involves reducing raindrop impact and wind action on soil particles by increasing soil cover through litter and pruning. Besides playing a vital role in reducing erosion and conservation of water resources, trees and shrubs fulfill another important function of stabilizing conservation structures, coupled with productive use of land, which they occupy.

Trees and shrubs are able to make effective use of pre-or-post-monsoon rains, which account for about one-fourths of the total southwest rains received. This is crucial in rainfed areas where every rain drops matters. Another heartening feature is that the tree as a main component of the agroforestry system can efficiently utilize 30 to 40 per cent of the rainwater due to their deep root system that goes into deeper soil layer where soils are predominantly light textured (Sharma and Gupta, 1997). Further, tree root systems help increase water infiltration into the soil and also intercept, absorb and recycle nutrients in the soil solution which otherwise have been lost in leaching, thereby making the nutrient cycle more closed. In following paragraphs an attempt has been made to highlight results of few studies on agroforestry which implies continuous presence of both annual and perennial groups of plants on the same site at the same time, like alley cropping, shelter belts and windbreaks, vegetative barrier, agrisilviculture, silvipasture, hortipasture system of land use.

Alley Cropping

Hedgerow intercropping is also known as alley cropping, the tree component is made up of a single or multiple rows of trees but essentially it has to be a dense hedgerow so as to be effective in reducing runoff, increasing infiltration and reducing soil loss through the barrier effect. Variables could be in terms of hedgerows, spacing between hedgerows or width of cropped alleys and management of pruning. *Leucaena* has been the most common and widely use species and to some extent *Gliricidia* has been used for hedgerow planting. Extensive studies on agronomic and fertility aspects of alley cropping systems have been made by Kang *et al.*, 1984, Kang and Duguma, 1985; Kang *et al.*, 1985 at the International Institute of Tropical Agriculture (IITA), Ibadan, Nigeria, but the effects of alley cropping systems on runoff and soil erosion, nutrient loss, moisture retention and availability to plants, etc. were not studied so far in very intensive manner. Lal (1989 a and b) reported the results of agroforestry systems and soil surface management of a tropical alfisol on water runoff, soil erosion, and nutrient loss, conducted at IITA, Ibadan, Nigeria, during 1982 to 1987. Soil moisture content in 0-5 cm layer in agroforestry system was generally higher than that in the control during both wet and dry seasons. Hedgerows essentially served as windbreaks and decreased soil moisture evaporation. Once hedgerows are established; they become extremely effective in reducing water runoff and controlling erosion. In contrast to runoff and erosion, losses of bases (Ca, K, Mg and Na) in runoff from agroforestry treatments were relatively high as compared to non-agroforestry systems, probably due to nutrient recycling by deep rooted perennial shrubs. Singh *et al.*, 1989 also stated that alley cropping is beneficial in terms of soil and water conservation with less runoff and soil loss with 3 m alleys than with 5 m alleys and root pruning or deep ploughing might be effective in reducing moisture competition.

Alley cropping with *Leucaena* is considered of prime importance in moisture conservation and efficient utilization thereof. Kanaujia *et al.* (2002) in their study on soil moisture content, extraction pattern and splash erosion under *Leucaena*–based alley cropping system carried out at Kanpur during 1993-95, found higher moisture content at sowing time in *Leucaena* planted at a close inter (4.0 m) and intra-row spacing (0.5 m) as compared to wider inter-row (6.0 m) and intra-row (1.0 and 1.5 m) spacing. Moisture availability under narrow inter row spacing (4.0 m) of *Leucaena* was higher at sowing time over wide interspacing of 6.0 m due to early interception of runoff and addition of higher amount of *Leucaena* lopping as organic mulch. At crop harvesting time, the trend of moisture content was reversed to that of sowing time due to higher demand of water by growing crops and densely planted *Leucaena*. In comparison to control, however, soil moisture regime under alley cropping was found high (except at harvesting of pearl millet 1994-95 due to low rainfall). Total moisture use in higher *Leucaena* plant density under narrow spacing (4.0 m) was at par or slightly higher than 6.0 m spacing mainly due to higher amount of *Leucaena* lopping as organic mulch, which obviously conserved higher amount of soil moisture for unwanted losses. Splash soil loss was also lower under closely spaced inter (4.0) and intra row spacing (0.5 m) of *Leucaena* over wider spacing because of higher biomass of *Leucaena* used as in-situ mulch which reduced the impact of falling rain drops.

On steep slopes dense multi-strata systems and alley cropping are especially useful in erosion control. The lopping and pruning from hedgerows could also provide mulch for preventing sheet erosion (Vergera, 1982). Alley cropping with mulching has been recommended as a means to tackle server erosion problems in Haiti (Zimmerman, 1986). The ameliorative effect of alley cropping on run off and soil erosion under maize-cowpea rotation is depicted in Table 10.1. Significant reduction in runoff and erosion was recorded in *Leucaena* alleys.

Vegetative Barriers

A number of vegetative materials have been tried for control of run-off and soil loss. The structures being pours, permit the run-off while retaining the soil and thus, overcome the problem of breaching. Vegetative barriers could include rows of perennial grasses, hedges, windbreaks and shelterbelts etc. on contours. In the NATP studies on soil and water conservation practices carried out by CRIDA, Hyderabad, it has been established that for all types of soil groups, grass cover is at least five times more effective than bare soil in the control of soil water loss. In a study, it was observed that runoff was completely arrested when groundnut was intercropped with pigeon pea mixed with pulses like cowpea and horse gram. Mishra *et al.* (1999) in their study on different vegetative barriers like Vetiver, Napier, Jatropha and Agave planted at 8 m interval in North Eastern Ghat zone of Orissa, they found that Vetiver to be the best among vegetative barriers and conserved maximum moisture. It reduced runoff by 20.3 per cent and soil loss by 51.4 per cent and increased soil moisture storage by 26.6 per cent over control. Krishnegowda *et al.* (1994) have summarized the research results on Khus (Vetiver) in Karnataka and reported that the grass is gaining popularity in watershed development programmes as a cost-effective measure for soil and water conservation. Vetiver hedge reduced soil loss from 11 t ha^{-1} in control to 2.5 t ha^{-1} in treated plots and reduced runoff from 26.3 per cent to 7.9 per cent respectively. Sur and Sandhu (1994) showed that Vegetative barriers reduced runoff and sediment loss in the order of Kanna (*Saccharum munja*) > Napier bajra hybrid (*Pennisetum purpureum*) > Vetiver > Bhabhar (*Eulaliopsis binata*) > without barrier.

Table 10.1: Effect of alley cropping on run ff and soil erosion under maize cowpea rotation.

Treatment	Run off		Soil Erosion
			(t ha^{-1} year^{-1})
	mm	*Per cent Rainfall*	
Plow till	232	17.1	14.90
No till	6	0.4	0.03
Leucaena			
4 m	10	0.7	0.20
2 m	13	1.0	0.10
Gliricidia			
4 m	20	1.5	1.70
2 m	28	2.8	3.30

Source: Lal, 1989.

Vetiveria was a component in most of the studies, however, the effectiveness of this species was not significantly better than the species of regional importance like *Eulaliopsis binata* in the Shiwaliks, *Cenchrus ciliaris* in the dry vertisols, *Pennisetum* in dry alfisols, *Panicum maximum* in the subhumid lower Western Himalayas. Vetiveria consistently failed to perform in the dry (semi arid) zones. Formation of gaps wider than 20cm. was the main cause against its successful performance as a barrier. For proper establishment of barrier hedge, a small bund of cross section 0.1 m2 on the downstream side has been found essential (Katyal *et al.*, 1994). Since vegetative barriers are economical to establish, easy to maintain, provide direct benefits like fodder etc. they could be a boon to marginal and small farmers. Patil *et al.* (1995) tried three vegetative barrier *viz.* khuss grass, subabul, weeping love grass and mechanical barrier of rubbles. Results revealed that barrier plot recorded minimum runoff and soil loss compared to control. When different barrier were compared, it was observed that the runoff and soil loss was minimum in rubble plot and maximum in plot with weeping love grass barrier (Table 10.2).

Table 10.2: Effect of barrier on runoff, soil loss and soil moisture content.

Barrier	*Runoff (mm)*	*Soil Loss (Kg plot^{-1})*
Rubble	61.1	821.7
Khuss grass	77.7	935.2
Subabul	79.3	877.2
Weeping love grass	93.3	1213.1
Control	132.6	1691.2

Silvipastoral System

This system is recommended for marginal soils. It involves integrating a tree component with a perennial legume or grass species as pasture. *Cenchrus ciliaris* and *Stylosanthes hamata* were extensively evaluated in different soil types and rainfall zones. Stylosanthes hamata is an improved pasture legume that can be raised on marginal lands and on field boundaries.

The potential of using woody perennials has often been emphasized for conservation as well as on hilly terrains; erosional losses were investigated in the western Himalayan region of India for a nine year period. The average annual rainfall was about 1000 mm and it caused 347 mm runoff and 39 Mg ha^{-1} soil loss per year for fallow plots. High density block plantation *Eucalyptus* or *Leucaena* almost completely checked runoff as well as soil loss. Pastoral or silvi-pastoral land uses also proved equally effective in controlling erosion. Integration of contour tree/hedge at 0.9 m vertical interval (22.5 m horizontal interval for 4 per cent slope) can reduce soil loss to a remarkable extent. The reduction in erosion was primarily due to the barrier effect of tree or hedgerows and micro terraces formed through sediment deposition along the contour barriers (Narain *et al.*, 1998). It was suggested that alley farming could effectively reduce erosional losses to an acceptable soil loss tolerance limit for medium to heavy textured and deep alluvial soil of the western Himalayan region.

Comparative study of some agroforestry system models for eroded lands indicated that in Shivaliks, *Eucalyptus-Eulaipsis binata* (bhabar grass) system was most effective (Table 10.3) in reducing the soil loss (0.07 t ha^{-1}) followed by *Acacia catechu*–napier grass (0.24 t ha^{-1}), *Leucaena*-napier grass (0.28 t ha^{-1}), Teak-*Leucaena*-bhabar (0.43 t ha^{-1}), *Eucalyptus-Leucaena*-turmeric (0.59 t ha^{-1}), Poplar-*Leucaena*–bhabar (1.54) as compared to arable crops (2.69 t ha^{-1}) and cultivation fallow (5.65 t ha^{-1}).

Studies on the conservative effect of different pasture management systems on runoff and soil loss on a watershed basis have shown that the losses in soil and soluble salts (N, P, K) may be minimized to a great extent in pasture based systems when compared to bare fields. Similarly, runoff losses were also reduced to a great extent in various pasture systems maintained at IGFRI, Jhansi (Table 10.4).

Shelterbelts and Windbreaks

Wind erosion can be controlled either through mechanical and chemical methods of sand stabilization, or through vegetative measures. Research and development activities at CAZRI since 1953 have perfected the methods of vegetative control of wind erosion (Bhimaya and Kaul 1960; Ganguli and Kaul 1969; and Kaul 1985). The major technological interventions in this context are stabilization of shifting sand dunes and shelterbelts plantation.

Table 10.3: Soil and nutrient loss from different land use systems at Shivaliks.

Land Use System	Study Years	Soil Loss (t ha⁻¹)	Runoff (Per cent of Rainfall)	Nutrient Loss (kg ha⁻¹)	
				N	K
Eucalyptus-bhabar grass	6	0.07	0.05	0.46	0.90
Acacia catechu-napier grass	3	0.25	2.00	6.97	0.52
Leucaena-napier grass	3	0.28	4.40	6.60	1.20
Teak-*Leucaena*-bhabar grass	3	0.43	3.30	2.08	0.55
Eucalyptus-Leucaena-turmeric	5	0.59	2.60	2.47	0.73
Poplar-*Leucaena*-bhabar	5	1.54	4.80	5.90	1.10
Sesamum-rape seed	3	2.69	20.50	42.50	3.00
Cultivated fallow	3	5.65	23.00	51.30	5.00

Source: Grewal, 1993.

Table 10.4: Average soil and nutrient loss under different pasture management systems.

Pasture Systems	Runoff (as per cent of rainfall)	Soil Loss (t ha⁻¹)	Nutrient Loss (kg ha⁻¹ year⁻¹)	
Bare	38.8	117.78	17.40	10.20
Natural Grassland	11.6	2.50	3.75	4.00
Improved Pasture	11.0	1.96	1.96	3.90
Sown Pasture	9.6	1.27	1.27	2.10
3-tier silvopasture	9.0	1.66	1.66	4.58

Experience suggests that across-the wind plantation of a 13 m wide tree belt, interspersed with 60 m wide grass belt, provides the best results. Establishment of micro-shelterbelts in a arable lands, by planting tall and fast growing plant species such as clusterbean on the windward side, and shorter crops such as vegetables in the leeward side of tall plants helped to increase the yield of lady's finger by 41 per cent and of cowpea by 21 per cent, over the control (Venkateswarlu 1993). In spite of the good results the community shelterbelts in arable lands are not very popular with the farmers, especially as the technique normally cut across their fields/holding boundaries. Therefore, of late, it has been suggested to plant trees on field bunds across the direction of wind (Ganguli and Kaul 1969; Venkateswarlu and Kar 1996; Singh *et al.*, 1987).

Shelterbelts reduce evaporation and moisture status of the sheltered field is generally higher by 2-4 per cent than unsheltered field (Gupta and Ramakrishna,

1988). As a result of these improvements in soil environment, a 20-30 per cent increase in yield of pearl millet is obtained. Besides reducing wind speed and evaporation, shelterbelts are useful in checking wind erosion and fertility loss from the agricultural fields. In western Rajasthan, three rows shelterbelt of *C.siamea – A. lebbek – C. siamea* is highly effective in reducing wind speed and loss of nutrients.

Agroforestry Potentials in Rainfed Areas

Agroforestry has high potential to simultaneously satisfy many important objectives *viz.* protecting and stabilizing the ecosystems, producing a high level of output of economic goods, providing stable employment, improve income and basic materials to rural population. Besides that agroforestry is capable to conserve natural resources through various systems under different agroclimatic regions

Agrisilviculture

In arid and semi-arid region, there are many indigenous systems of trees on cropland or on pastures, a feature of which is that the tree component nearly always survives periods of drought. Water conservation beneath trees, leading to higher pasture production, is an element in silvipastoral systems. *Prosopis cineraria* – An N-fixer is an important MPT from the region and source of animal feed, fuel and timber. It has a very deep tap root and it can be lopped at a young age (about 8 years) (Singh, 1987). In agroforestry systems with relatively low densities of *P. cineraria* (about 120 trees ha^{-1}, depending on soil type and rainfall), it is usually intercropped with millet and legumes. Improved soil fertility and higher moisture content have been found in *P. cineraria* inter-cropped systems, along with higher grain yield and forage biomass production (Singh, 1987).

In arid and semi-arid environments, agroforestry systems help to provide greater insurance against weather abnormalities. MPTS such as species of *Acacia, Prosopis* and *Casuarina* are commonly used in agroforestry combined with grain crops. The large tap root system in trees indigenous to dry regions indicates an ecological adaptation to extraction of water from deep soil layers during dry periods. Such roots may penetrate well into the spaces between stones or weathered rock or along joints (Dhyani *et al.*, 1996). This allows the tree to survive through long dry seasons. However, there is a striking disparity between the occurrence of higher soil water content beneath indigenous trees in dry regions and reports of water depletion by planted multipurpose trees. The much faster growth of the latter is the probable explanation (Young, 1997).

Silvipastoral System

Forage production under silvipastoral system is directly influenced by its proper initial establishment, subsequent management, edaphic and climatic conditions, tree density, canopy development, etc. The literature available on forage production under different silvipastoral systems in different agroclimatic conditions was generally on the basis of 4 to 6 years studies (Singh and Puri, 1975; Agrawal *et al.*, 1976; Desai, 1978; Pradhan and Vaswa, 1978; Deb Roy *et al.*, 1980; Humpaih, 1981; Muthana *et al.*, 1985; Singh and Roy, 1991; Bhattacharya and Sharma, 1993; Rao and Usman, 1994;

Sharma *et al.*, 1994; Hazra *et al.*, 1994; Rana *et al.*, 1995; Sharma *et al.*, 1996; Dogra *et al.*, 1997; Arya *et al.*, 1998, Rai, 1999a and b Rai *et al.*, 2000 a and b, 2001, 2002 etc.). In general, through lopping/pruning, tree contributes meager in the total forage production obtained from silvipastoral systems (Pasture+trees). A long term studies conducted on silvipastoral systems at Jhansi (U.P.) and Datia (M.P.) in red gravelly soils with five tree species at each location (*Albizia amara, Albizia lebbeck, Acacia tortilis, Hardwickia binata and Leucaena leucocephala*) at Jhansi and (*A. amara, A. lebbek, A. tortilis, Dalbergia sissoo and L. leucocephala*) at Datia under 3 spacing (4 x 3, 4 x 4 and 4 x 6 m) revealed that tree did not have any adverse effect on forage production up to 11-12 years at any spacing (IDRC Reports, 1982-83 to 1993-94). Thus, the results showed that trees may be grown at the rate of 825 plants ha^{-1} up to 11-12 years without significant reduction in understorey biomass.

In arid condition, silvipasture is the most important land use system. It has been observed that in addition to indigenous species like *Prosopis cineraria and Zizyphus nummularia*, the introduced species such as *A. tortilis, D. cinerea, Colophospermum mopane* showed very good performance in Silvipastoral system (Ahuja, 1980). Production from lower storey crops under 14-18 years old plantation of 4 desert tees (*P. cineraria, Tecomella undulata, A. lebbek and Acacia senegal*) ranged from 0-6 to 1.5 t ha^{-1} year^{-1} (Ahuja, 1980). Productivity of arid land can be increased 2-3 times by replacing natural grass cover with C. ciliaris and introduction of top feed species like *Z. nummularia and Grewia tenax* under silvipastoral system (Sharma *et al.*, 1994).

In another studies of silvipastoral system conducted at C.S.W.C.R. and T.I., Regional Research Station, Bikaner (arid climate) having *P. cineraria, Z. mauritiana, A. tortilis* and *C. mopane* as a tree components in association with *Lasirus sindicus* (Sewan grass) pasture, the total dry matter and Sewan grass yields obtained ranged from 3.50 to 3.97 t ha^{-1} and 2.64 to 3.10 t ha^{-1}, respectively (Anonymous, 1998). However, silvipasture had to be re-established after some years due to poor survivability of tree species.

A thirteen years studies on degraded lands with five tree species and three densities under silvipastoral systems revealed that the average total biomass production of 6.55 to 8.53 t ha^{-1} was obtained in association with *L. leucocephala* followed by *A. amara*. Average total biomass production of 6.91 t ha^{-1} was recorded when planting was done at 4 x 3 m spacing followed by 4 x 4 m and minimum with 4x6 m spacing. Studies further revealed that for higher biomass production, trees should be planted at the density of 825 plants ha^{-1} (4m x 3m) and maintained at least up to 12-13 years after that if understorey production starts decreasing felling of trees should be done. If harvesting of trees was done at this age, at least 2 to 3 t ha^{-1} year^{-1} biomass may be obtained. Thus, the total biomass production from silvipastoral system can be obtained by about two times higher than that pasture alone.

On the basis of more than 20 years studies in shallow red gravelly soils under semi arid condition at Jhansi, Pathak *et al.* (1996) reported that the degraded lands producing hardly up to 1 t ha^{-1}year^{-1} at a 10 years rotation through silvipastoral systems. Besides yield improvement by 8 to 10 times, the quality of mixed forage has also improved by 6 to 7 times.

A study conducted at NRCAF, Jhansi to compare the biomass yield from silvipastoral system and natural grassland for 8 years revealed that on an average the total biomass yield was 12.62 t ha^{-1}year^{-1} under silvipastoral system which was about 4 times higher than yield obtained from natural grassland (3.16 t ha^{-1}year^{-1}). These results showed that it is possible to get more than 12 t/ha/year biomass through established silvipasture on the land which is producing only 3 t ha^{-1}year^{-1} biomass through natural vegetation.

Agrihorticulture System

Agrihorticulture system is an alternative land use system that integrates the cultivation of arable crops and fruit trees. Because of high value and demand of fruits, agrihorticulture occupies a prime position in the farming communities. Growing of mango, guava, aonla, pomegranate, ber, bael, citrus with agricultural crops either in irrigated or rainfed conditions are some examples of agrihorticulture. Normally, mango, guava and citrus based agrihorticulture systems are recommended for better land with irrigation facilities. But other fruit trees those required less water or able to tolerate drought are very much suitable for rainfed areas with rainy season crops. Results gathered on economics of agrihorticulture to date show that the benefit-cost ratio with agrihorticulture system was 2.5 as against 1.50 with rainfed arable crops (Ram Newaj *et al.*, 2004). Similarly, several studies showed that integrated cultivation of aonla with kharif pulses gave higher monetary returns from agrihorticulture system compared with mono-crops (Solanki and Ram Newaj, 2002). Besides higher monetary returns, inclusion of fruit trees with crops increases the soil cover, which reduces the runoff and soil losses than mono-cropped system. The adoption rate of the "agrihorticulture" system is more rapid than other systems on account of risk distribution. The tree component is able to absorb weather related aberrations more efficiently. This approach lessens the financial imbalances during gestation period by regulating the income through intercropped annuals when the trees are too young to yield beneficial produce (Solanki *et al.*, 1999).

National Research Centre for Agroforestry has been working on agrihorticulture system since 1996 on *in-situ* moisture conservation in very poor soil which having very shallow depth and low moisture holding capacity known as rakar soil of the Bundelkhand regions. For conserving soil moisture, the aonla (variety NA 7) was planted with different planting techniques like sunken method of planting, stone mulching and sunken method of planting associated with deep tillage followed by kharif crop to retain residual moisture and increase the water infiltration into deeper soil layer. For comparing the benefit of *in-situ* water conservation, the traditional method of planting (farmer's practice) was also included in the study. In kharif season, greengram was grown as intercrop and after picking the pods, the crop was ploughed for green manuring. The aonla was planted at 10 m x 10 m spacing. The results obtained from this study were very encouraging and moisture conservation techniques always gave higher yield of fruit, net income, benefit:cost ratio and conserved 20 per cent higher moisture than traditional method of planting. This system can yield 3.42 t ha^{-1} fruit, 0.012 t ha^{-1} green gram per year after 7 years. In this way the system can generate net income Rs.37252 from 7th year. Although income obtained from the system during initial year may not be positive but overall, the system can give the benefit of Rs. 1.8 on per rupee expenditure when fruit tree start fruiting

Agrisilvihorticulture System

Agrisilvihorticulture system integrate the cultivation of arable crops, fruit trees and fast growing multipurpose trees preferably nitrogen fixing trees which may be used for fuel and fodder purpose with frequent harvesting so that competition between fruit trees and crops may be minimized. NRC for agroforestry initiated a field trial during 1989 with 4 varieties of aonla *viz.* Chakaiya, Kanchan, Krishna and NA 7 as fruit trees, leucaena as multipurpose tree and blackgram as intercrop in rainfed areas. The leucaena was planted on both sides of the fruit trees at 2m distance. Soil of the experimental site was mixed red and black, representing characteristics of light soil with low moisture holding capacity. It represented soil of bundelkhand region, having pH 7.9 EC 0.16 mmhos cm^{-1}, ESP 0.83 me/100g, organic carbon 0.32 per cent, nitrogen 161.7 kg ha^{-1}, phosphorus 13.2 kg ha^{-1} and potassium 120.6 kg ha^{-1}. The aonla was planted at 10m x 6m and 5m x 6m spacing but 10m x 6m spacing was proved an ideal spacing among these and it was considered for calculating the yield and economics of the system

Fruit bearing in aonla had started 4 years after plantation. The fruit yield during initial years was low and it increased with subsequent increase in age and growth of fruit trees. In age of 6 year, on an average the fruit yield from a plant was up to 409 kg ha^{-1} (Table 10.5). The fruit yield of aonla varied year to year which clearly indicated that in alternate year the fruit yield was higher and it also has a relation with rainfall received during preceding year. Besides fruit yield of aonla, 256 kg grain was obtained form blackgram every year (Table 10.5).

Introduction of leucaena in the system provide organic matter in the form of leaf litter and it also fixed atmospheric nitrogen in the soil. *Leucaena* was cut twice a year and it provided on an average 1325kg fuel wood and 799kg leaves ha^{-1} every year (Table 10.5). The leaves of leucaena were utilized as mulch to minimize moisture loss from soil during summer, which is very essential in alfisols in rainfed areas. Besides fuel and fodder yield from leucaena, it also help in improving soil fertility. In this system, organic carbon was increased after 9 years up to 28.1 per cent from original value (0.32 per cent).

Based on input required for cultivation of aonla in agri-silvi-horticulture system, the cost of each and every input was included to analyze the economics of the system. The cost of cultivation in first year which includes planting of fruit trees and leucaena as multipurpose trees and cultivation of crop was Rs. 8,666 but during next year the cost of cultivation was reduced and it was again increased with subsequent increase in the cost of input during different years. The gross income from the system was less during initial year but when fruiting started in aonla, the gross income was increased and it gone up to Rs. 60,712 in age of 13 years (Table 10.5). Similarly the net income was positive in all the years except first, third and fifth years when aonla had no fruiting/or less fruiting. In age of 13 years, the B: C ratio from the system was 3.28 and on discounted rate it was 2.61 which indicated that aonla based agroforestry system is a profitable enterprise in marginal lands under rainfed conditions.

Table 10.5: Economics of the agri-silvi-horticulture (aonla+ leucaena + blackgram) system.

Year	Cost of Cultivation (Rs.ha⁻¹)	Gross income (Rs.ha⁻¹)	Net income (Rs.ha⁻¹)
89-90	8,666	3,450	-5,216
90-91	4,613	5,016	403
91-92	5,075	3,604	-1,471
92-93	6,018	7,774	1,756
93-94	6,966	14,051	7,085
94-95	6,934	4,432	-2,502
95-96	8,170	27,343	19,173
96-97	7,975	28,686	20,711
97-98	8,302	31,183	22,881
98-99	8,640	41,776	33,136
99-2000	9,668	59,980	50,512
2000-01	8,952	40,185	31,233
2001-02	8,975	27,496	18,521
2002-03	9,200	60,712	51,512

Simple benefit: cost ratio =3.28; Discounted benefit: cost ratio @ 10 per cent =2.61.

Source: Ram Newaj (unpublished data).

Hortipastoral System

A combination of fruit trees and pasture species, commonly known as "hortipastoral system" it satisfy human needs in several ways and alleviate cattle hunger. A hortipastoral system comprising of hardy fruit trees is advocated for land capability class (LCC) III and IV. To improve wasteland/degraded lands through hortipastoral system, the first step is to protect from biotic interference followed by selection of suitable top-feed species which have fast growth and good coppicing ability and are highly palatable.

Selection of fruit species for hortipastoral system under semi-arid environment is quite difficult task. Prior to selection of fruit species, it is pre-requisite to select a particular variety for that region. Singh and Osman (1995) reported that such tree should be chosen which are complementary or less competitive for understorey plants. Further, slow growing and late bearing fruit tree species *e.g.* tamarind (*Tamarindus indica*), wood apple (*Limonia acidissima*), jamun (*Syzygium cuminii*), aonla desi (*Emblica officinalis*), beal (*Aegle marmelos*) are highly suitable for dryland areas. In a continuous grazing system, custard apple could be an ideal option as it is not browsed by live stock and hardy too. Fruit like ber will grow very well in combination with pasture, as the foliage is very nutritious and palatable. Vashishtha (1991) recommended that the fruit and grass/legume should be selected on the basis of rainfall pattern and some suitable trees and grass/legumes are given in table 6.

Involvement of fruit trees in farming system reduces risk (Dayal *et al.*, 1996). Korwar *et al.* (1988) reported that the fruit trees are very hardy, deep rooted and tolerant to abnormal monsoon *e.g.* early or late onset, intermittent dry spell, early withdrawl, uneven distribution of rainfall etc. better than short duration field crops. These trees yield at least something in drought year when seasonal crop totally fail. Thus, it results in stability of income and brings ready cash to farmer. The fruit viz; aonla, bael, custard apple, ber pomegranate, guava, jamun, gonad etc. can be grown successfully in semi-arid condition (Vashishtha, 1991).

Table 10.6: Suitable fruit crops for different rainfall zones.

Rainfall (mm) Zone	Fruit Crop	Grass/Legume
Less than 350	Ber (*Zizyphus mauritiana*)	Buffel grass (*Cenchrus ciliaris*)
	Gonda (*Cordia myxa*)	Bird wood grass (*C. setigerus*)
	Ker (*Capparia deciduas*)	Sain (*Sehima nervosum*)
	Pilu (*Salvadora oleoides*)	Stylo (*Stylosanthes scabra*)
		Butterfly pea (*Clitoria ternatea*)
		Ban kulthi (*Atylosia scarabaeoides*)
350-500	Ber (*Zizyphus mauritiana*)	Blue panic (*Panicum antidotale*)
	Aonla (*Emblica officinalis*)	Marvel (*Dichanthium annulatum*)
	Khirni (*Manilkara hexandra*)	Dinanathgrass (*Paspalum pedicellatum*)
	Jamun (*Syzygium cuminii*)	Bahia grass (*Paspalum notatum*)
	Mulberry (*Morus alba*)	Sabigrass (*Urochloa mosmabicensis*)
		Stylo (*Stylosanthes hamata*)
500-700	Bael (*Aegle marmelos*)	Rhodes grass (*Chloris gayana*)
	Mango (*Mangifera indica*)	Stylo (*Stylosanthes hamata*)
	Custard apple (*Annona spp.*)	Siratro (*Macroptilium atropurpureum*)
	Sour lime (*Citrus aurantifolia*)	
	Aonla (*Emblica officinalis*)	
More than 700	Mango (*Mangifera indica*)	Napiergrass (*Pannisetum purpureum*)
	Pomegranate (*Punica granatum*)	Marvel (*Dichanthium annulatum*)
	Guava (*Psidium guajava*)	
	Jackfruit (*Artocarpus heterophyllus*)	
	Wood apple (*Limonia acidissima*)	
	Tamarind (*Tamarindus indica*)	
	Aonla (*Emblica officinalis*)	

Source: Vashishtha, 1991.

Studies carried out under the All India Co-ordinated Research Project for Dry land Agriculture (AICRPDA) during 1975-83, revealed that only a few species of grasses and legumes are suitable for dryland areas. Different grasses from genera *Dicanthium, Cenchrus, Lasiurus, Chloris, Urochola, Panicum, Pennisetum, Sehima, Chrysopogan, Paspalum, Digitaria, Heteropogan,* have been evaluated. Of these, *Dicanthium, Sehima* and *Lasiurus* were found to be ideal for severe drought prone areas. *Cenchrus ciliaris, Panicum maximum (Guinea mucunae),* and *Urochloa* were found ideal for moderate drought prone areas.

Establishment of fruit trees in drylands requires proper microsite improvement and timely weeding during initial period. Studies at CRIDA have indicated that ring weeding and *in situ* moisture conservation are essential to improve the survival of fruit tree seedling in drylands (Korwar *et al.,* 1997). In another study at CRIDA, survival was found to be highly related to the control of native grasses. Custard apple and jamun had high survival when native grasses were controlled by ring weeding and ploughing once during rainy season.

Preliminary studies on survival and growth of 12 fruit trees under natural grassland condition revealed that performance of *Emblica officinalis, Cordia myxa, Morus rubra, Zizyphus mauritiana, Tamarindus indica* performed better compared to other fruit trees which may be due to their ability to withstand the comptetion with grasses growing around the fruit plants in hortipastoral system (Table 10.7). The dry forage yield ranged from 3.51 to 5.49 t ha^{-1} during 3rd year of the study.

Table 10.7: Survival, growth parameters of fruit trees and dry forage yield under natural grassland.

Fruit trees species	Survival (per cent)	Growth Parameters (2.5 year age)		Dry Forage Yield
		Plant Height (m)	Collar Diameter (cm)	(t ha^{-1}) in 3rd year
Emblica officinalis	93.3	2.22	5.63	3.51
Buchanania lanzan	0.0	0.24	0.52	3.89
Annona squmosa	6.7	0.59	1.02	4.92
Aegle marmelos	73.3	0.86	1.97	4.47
Cordia myxa	100.0	2.03	5.59	3.91
Morus rubra	86.7	2.73	5.14	4.51
Tamarindus indica	86.7	1.38	3.73	4.31
Punica granatum	80.0	1.34	2.44	5.49
Achras zapota	20.0	0.53	1.02	5.68
Manilkara hexanda	80.0	0.83	2.17	4.79
Limonia acidissima	73.3	0.60	1.23	4.62
Zizyphus mauritiana	86.7	1.69	2.59	4.68

Source: Shukla *et al.,* 1998.

Studies conducted at Indian Grassland and Fodder Research Institute, Jhansi by Sharma (1996) showed no effect of growing grass on growth of ber trees. Growth parameters namely plant height, stem diameter and tree crown diameter of ber (*Zizyphus mauritiana* cv Gola) were maximum with grass (*Cenchrus ciliaris*) as compared to other treatment combinations although they were statistically not significant. Fruit production increased in association with legume stylo (*Stylosanthes hamata*) compared to control and decreased in association with grass (Table 10.8).

Table 10.8: Growth and production of ber based hortipastoral system.

Treatment Combinations	Tree growth parameter			Fruit Yield (kg tree⁻¹)	Pasture (DM t ha⁻¹)
	Plant Height (cm)	CD (cm)	Crown Diameter (cm)		
Tree alone	321	7.28	332	20.48	-
Tree+grass	354	8.48	386	17.74	5.63
Tree+legume	340	7.69	372	25.35	4.00
Tree+Grass+Legume	314	7.56	374	22.45	4.32
CD at per cent	NS	NS	NS	3.06	NS

Source: Sharma, 1996.

Prasad *et al.* (1997) reported highest average girth of aonla 38.4 cm in association with *Chrysopogan fulvus* and 32.4 cm with napier grass when compared to 37.7 cm of pure aonla. Height of aonla was maximum (629 cm) in pure situation and minimum in association with napier. Fruit yield and number of fruit per tree were also minimum (518 fruit tree⁻¹) in association with napier and were maximum when growth with *Chrysopogan* (1935 fruit tree⁻¹). *Chrysopogan* and napier grass when grown in combination with aonla gave an average yield of 8.1 and 3.8 t ha⁻¹ (air dry weight) respectively.

Effect of nitrogen on productivity of aonla based hortipastoral system revealed that application of 60 kg N ha⁻¹ to Dichanthium pasture produced significantly higher mean yield of dry forage (6.15 t ha⁻¹), aonla fruit yield (5.82 t ha⁻¹) and benefit cost ratio of 2.35 under rainfed condition compared to 0, 20 and 40 kg ha⁻¹ (Kumar *et al.*, 2002).

Studies on effect of 3 levels of pruning (severe : 20 cm length of secondary branches intact, medium : 40 cm length of secondary branches and light : 60 cm length of secondary branches) of 10 years old ber plantation (6m x 6m spacing) and 4 pasture combinations (Guinea grass + stylo, Dinanath grass + stylo, Guinea grass + Dinanath grass + stylo and natural grasses) revealed that medium pruning gave on an average significantly higher fruit yield (29.33 kg tree⁻¹) followed by light (25.9 kg tree⁻¹) and severe pruning (19.95 kg tree⁻¹) while fuel wood was maximum in light pruning (34.45 kg tree⁻¹) followed by medium and severe pruning. The pasture yield was found maximum in severe pruning (4.29 t ha⁻¹) followed by medium and light pruning. Out of 4 pastures, Guinea grass + stylo showed maximum yield of 5.87 t ha⁻¹ and lowest yield of 1.29 t ha⁻¹ in natural pasture (Kumar *et al.*, 2004).

In a study at CRIDA (1996), a survival of 95 per cent was recorded in guava after six years of planting by controlling the grass in the tree basins (2 m diameter) and pot watering of plants during dry period. Yield reduction of stylo was found to be less under widely spaced plats (8 x 5 m) compared to closer spacing (5 x 5 m), indicating the necessity of wider spacing of fruit trees when grown with stylo. The fruit yield increased significantly with closer spacing and also when forage was not planted in association (Table 10.9).

Table 10.9: Fresh yield of forage and fruit in Guava based hortipasture system.

Spacing (m)	Forage Yield (t ha⁻¹)		Fruit Yield (kg plant⁻¹)
	Stylo Legume	Buffel Legume	
5 x 5	5.22	2.45	95.4
8 x 5	6.56	2.14	99.7
Control	8.76	2.55	-
Mean	6.84	2.38	97.5
SE	0.29	0.55	3.09
CD (0.05)	0.88	NS	NS

Source: Osman and Rao, 1999.

Conclusion

In semiarid and arid areas, water is precious input and it needs to conserve/ harvest very efficiently either through biological or mechanical method. Biological method is a cheapest and easier to conserve our natural resources like soil, water and vegetation. Among biological method, agroforestry is most efficient land use system by which we can conserve our natural resources and can meet out our increasing demands of fuel, fodder, timber etc. In semiarid and arid areas, proper care should be taken in selection of fuel/timber/fruit trees and in case of fruit tree, preferably it should be quick yielding like aonla and ber. Hardy, slow growing, widely spread fruit plants are best suited for hortipastoral system. Stylo legume may be preferred over buffel grass because of its ability to improve soil physico-chemical properties and enrich diet of livestock deficient in crude protein content. Trees and grasses should be selected according to rainfall and soil condition of the particular areas.

References

Agrawal, R.K., Gupta, J.P., Saxena, S.K. and Muthana, K.D.1976. Studies on soil physico-chemicals and ecological changes under twelve years old five desert tree species of western Rajasthan. *Indian Forester* 102: 863-872.

Agroforestry Systems, 4: 255-268.

Ahuja, L.D. 1980. Grass production under Khejri. In: Khejri (*Prosopis cineraria*) in Indian Desert- its role in agroforestry (Eds. H.S. Mann and S.K. Saxena) Monograph No. 11, CAZRI, Jodhpur pp. 28-30.

Anonymous, 1998. 35 years of Research. Central Sheep and Wood Research Institute, Avikanagar (Rajasthan), pp.1-128.

Arya, R.L., Singh, A., Yadav, R.B., Singh, S.lP., Gangwar, M.K. and Lal, Banwari 1998. Early performance of trees and grasses under silvipastoral systems in salt affected soils of indo Gangetic plains of Uttar Pradesh. *Range Management and Agroforestry* 19: 53-57.

Bhattacharya, N.K. and Sharma, K 1993. Silvipastoral system in relation to goat farming. In: Agroforestry in 2000 AD for the Semi arid and Arid tropics Extended Abstract (Late arrivals). (Eds. A.S. Gill, R. Deb Roy and A.K. Bisaria), NRCAF, Jhansi.

Bhimaya, C.P. and Kaul, R.N. 1960. Some affoerstation problems and research needs in relation to erosion control in arid and semi arid parts of Rajasthan. *Indian Forester* 87:354-367.

Dayal, S.K.N., Grewal, S.S. and Singh, S.C. 1996. An agri-silvi-horticultural system to optimize production and cash return for Sivalik foot hills. *Indian J. of Soil Conservation* 24 (2): 150-155.

Deb Roy, R., Patil, B.D., Pathak, P.S. and Gupta, S.K. 1980. Forage production of Cenchrus ciliaris and Cenchrus setigerus under silvipastoral system. *Indian J. Range Mgmt.* 1: 113-120.

Desai, S.N. 1978. Silvipastoral system of production and its scope in dry areas of Maharashtra. Proceeding Summer Institute on Silvipastoral Production held at IGFRI, Jhansi.

Dhyani, S. K., Puri, D.N. and Narain, P. 1996. Biomass production and rooting behaviour of *Eucalyptus tereticornis* . On deep soils and riverbed bouldery lands of Doon Valley. *Indian Forester* 122 (2): 128-136.

Dogra, K., Katoch, B.S., Sood, B.R. and Singh, Gurudev 1997. Production potential and quality of silvi-herbage systems vis-à-vis natural grasslands in the humid sub-tropics of Himachal Pradesh. *Range Mgmt. and Agroforestry* 18: 165-169.

Ganguli, J.K. and Kaul, R.N. 1969. Wind Erosion Control. ICAR Technical Bulletine (Agric) No. 20, ICAR, New Delhi-57p.

Grewal, S. S. 1993. Agroforestry systems for soil and water conservation in Shivaliks. In: *Proceedings Agroforestry in 2000 AD for the Semiarid and Arid Tropics* (eds. A. S. Gill, R. Deb Roy and A. K. Bisaria). NRCAF, Jhansi: 82-85.

Gupta, J.P. and Ramakrishna, Y.S. 1988. Role of shelter belts in checking wind erosion and increasing crop production in sandy wastelands of western Rajasthan. In: *Wastelands Develop.*

Hazra, C.R., Singh,D.P. and Kaul, R.N. 1994. Greening of Common Lands in Jhansi through Village Resource Development- a case study. National Aforestation and Eco-development Board, New Delhi.

Humpaih, R. 1981. Improved forage forestry for semi arid regions of India with special reference to Bundelkhand region. *Indian J. Range Mgmt.* 2: 81-85.

IDRC Report 1982-83 to 1993-94. Annual Report of IDRC Silvipastoral Operational Project. IGFRI, Jhansi.

Kanaujia, V.K., Uttam, S.K. and Kaushal Kumar 2000. Soil moisture content, extraction pattern and splash erosion under leucaena based alley cropping system in eroded alluvial soil of central Uttar Pradesh. *Indian Journal Agroforestry* 4(2): 113-118.

Kang, B.T. and Duguma, B. 1985. Nitrogen management in alley cropping systems. In: Kang, B.T. Van der Heida, J. (Eds.). *Nitrogen Management in Farming systems in Humid and Sub-Humid Tropics.* Institute of Soil fertility, Haren, Netherlands, pp. 269-284.

Kang, B.T., Grimme, H., and Lawson, T.L. 1985. Alley cropping sequentially cropped maize and cowpea with *Leucaena* on a sandy soil in Southern Nigeria. *Plant and Soil* 85: 267-276.

Kang, B.T., Wilson, G.F., and Lawson, T.L. 1984. Alley Cropping: a stable alternative to shifting cultivation IITA, Ibadan, Nigeria.

Katyal, J.C., Sharma, S., Das, S.K. and Mishra, P.K. 1994. Moisture conservation and rain water management in red soil regions. *Indian Journal of Soil Conservation* 22 (1-2): 15-25.

Kaul, R.N. 1985. Afforestation of sand dune areas. In Sand Dune Stabilization, Shelterbelts and Afforestation in Dr Zone. FAO Conservation Guide 10. Rome Italy, FAO.pp75-85.

Korwar, G.R., Mohd. Osman, Tamar, D.S., Singh, R.P. 1988. Extension Bulletin No. 4 CRIDA, Hyderabad.

Korwar, G.R., Rao, J.V., Osman, M. and Pratibha, G. 1997. Agroforestry work at CRIDA-A Brief Note. Paper presented at the workshop of AICRP on Agroforestry, held at ANGRAU Rajendranagar, Hyderabad, Jan. 28-30, 1997.

Krishnegowda, K.T., Dabnay, S.M.Dunn, G.H., Meyer, L.D. Harmon, W.C., and Alniso, C.V. 1994. Investigation on vegetative barriers for soil and moisture conservation. *Abstracts of Papers, 8th ISCO conference*, New Delhi, pp. 372.

Kumar, Sunil, Kumar, Sudhir, Baig, M. G., Choubey, B. K. and Sharma, S. K. 2002. Effect of nitrogen on productivity of aonla based hortipastoral system. *India J. Agroforestry* 4 (2): 94-97.

Kumar, Sunil, Ram, S. N., Baig, M. J. and Roy M. M. 2004. Productivity of ber based hortipastoral system on red alfisol : Effect of pruning intensities and pasture combinations. *India J. Agroforestry* 6 (1): 23-26.

Lal, R. 1989 b. Agroforestry systems and soil surface management of a tropical alfisol: II Water runoff, soil erosion and nutrient loss. *Agroforestry Systems* 8: 97-111.

Lal, R. 1989a. Agroforestry systems and soil surface management of a tropical alfisol: I. Soil moisture and crop yields. *Agroforestry Systems* 8: 7-29.

Mishra, P.K., Shrinivas Sharma, Rao K.V., and Singh, H.P. 1999. Potentiality of Water harvesting and use at microlevel in semi arid Alfisols - A case study. *Proceeding of National Seminar* on Water Resource Development and Management of sustainable crop production, WALMTARI, Hyderabad.

Muthana, K.D., Sharma, S.K. and Harsh, L.N. 1985. Studies on silvipastoral system in Arid Zone. *My Forest* 21: 233-338.

Narain, P., Singh, R.K., Sindhwal, N.S., and Joshie, P. 1998. Agroforestry for soil and water conservation in the Western Himalayan Valley Region of India. I Runoff, soil and nutrient losses. *Agroforestry Systems* 39: 175-189.

Osman, M. and Rao, J.V. 1999. Alternate land use: Hortipastoral system. In : *Fifty years of dryland agricultural research in India.* (eds. H.P. Singh, Y.S. Ramakrishna, K.L. Sharma and B. Venkateswarlu), CRIDA, Hyderabad. pp 485-495.

Pathak, P.S., Gupta,S.K. and Singh, P. 1996. IGFRI Approaches Rehabilitation of Degraded Lands. IGFRI, Jhansi, pp.1-23.

Patil, Y.M., Belagaumi, M.I, Maurya, N.L., Kubasad, V.S., and Mansur, C.P. 1995. Impact of mechanical and vegetative barriers on soil and moisture conservation. *Indian J. Soil. Cons.* 23(3):254-255.

Pradhan, I,P. and Vaswa, S.S. 1978. Utilization of class VI and VII lands for fuel and fodder. Annual Report, CSWCR and TI, Dehradun, pp. 60-61.

Prasad, R. Saha, B., Sarma, J.S. and Agnihotri, Y. 1997. Development of hortipastoral land use system in degraded land. *Annual report*, Central Soil and Water Conservation Research and Training Institute, Dehradun, India. 41 p.

Rai, P. 1999a. Silvipastoral system – Growing greener and increasing the production of wastelands. *Wastelands News* XIV (3) : 48 -50.

Rai.P. 1999b. Comparative growth and production of MPTS at 8 years establishment under natural grassland in medium black soils in semi arid condition at Jhansi. *Range. Mgnt. and Agroforestry* 20(1): 94-97.

Rai, P., Solanki, K. R. and Singh, U.P. 2000a. Growth and biomass production of multipurpose tree species in natural grassland under semi arid conditions. *Indian J. Agroforestry* 2 (1): 101-103 .

Rai, P. Rao, G. R., and Solanki, K. R. 2000b. Effect of multipurpose tree species on composition, dominance, yield and crude protein content of forage in natural grassland. *Indian J. Forestry* 23 (4): 380-385.

Rai, P., Yadava, R. S., Solanki, K. R., Rao G. R. and Singh Rajendra, 2001. Growth and pruned production of multipurpose tree species in silvipastoral system on degraded lands in semiarid region of Uttar Pradesh, *India. Forests, Trees and Livelihood* . 11: 347-364.

Rai, P., Dadhwal, K. S., Solanki K. R and Tiwari, R.. 2002. Role of agroforestry in disaster management. *Vigyan Bharti.* 8 (1): 103-110.

Ram Newaj, Rai, P. and Ajit 2004. Effect of moisture conservation techniques on growth and yield of aonla in agrihorticulture system under rainfed conditions. *Multipurpose Trees in the Tropics: Assessment, Growth and Management* International Conference at Arid Forest Research Institute p. 195.

Rana, B.S., Saxena, A.K., Rao, O.P. and Singh, V.P. 1995. Para grass for silvipastoral system on saline sodic lands. *Agfororestry Newsletter* 7: 6-8.

Rao, J.P. and Usman, M. 1994. Studies on silvipastoral systems in non-arable drylands. In: Agrororestry Systems for Degraded Lands. (Eds. P. Singh, P.S. Pathak and M.M. Roy). Oxford and IBH Pub. Co. Pvt. Ltd., New Delhi, pp. 755-760.

Sharma, B.M. and Gupta, J.P., 1997. Potential role of agroforestry zone in 2000 A.D In: *Agroforestry for Sustained Productivity*, Gupta, J.P. and Sharma, B.M. (Eds.), Scientific Publishers, Jodhpur, 198p.

Sharma, S.K. 1996. Growth and productivity of fruit crop in association with grasses and legumes. Annual Report, Indian Grassland and Fodder Research Institute, Jhansi: 39-40pp.

Sharma, S.K., Singh, R.S., Tiwari, J.C. and Burman,U. 1994. Silvipastoral studies in arid and semi arid degraded lands of western Rajasthan. In-Agroforestry Systems for Degraded Lands (Eds. P. Singh, P.S. Pathak and M.M. Roy). Oxford and IBH Pub. Co. Pvt. Ltd., New Delhi pp. 749-754.

Sharma,S.K.,Dutta, B.K. and Tiwari, J.C. 1996. Prosopis cineraria (L.) Druce in silvipastoral system in arid regions of western Rajasthan. *Range Mgmt. and Agroforestry* 17: 81-85.

Shukla, S. K., Rai, P. and Rao, G. R. 1998. Preliminary studies on survival and growth of fruit trees and forage production in rangelands. *Range. Mgmt. and Agroforestry* 19 (1) : 93-96.

Singh, G.B. 1987. Agroforestry in the Indian sub-continent: past, present, and future. In (HA Steppler and P.K.R. Nair eds.). *Agroforestry: A Decade of Development*. ICRAF, Kenya, pp. 117-138.

Singh, J.P. and Puri, D.N. 1975. Economic utilization of ravines for fuel cum fodder. Annual Report, CSWCR and TI, Dehradun. pp. 39-40.

Singh, P. and Roy, M.M. 1991. Forage production through agroforestry constraints and priorities. *Range Mgmt. and Agroforestry* 12: 169-178.

Singh, R.P. and Osman, M. 1995. Alternate land use systems for drylands. Pages 375-398 in Sustainable Development of Dryland Agriculture in India (Singh, R.P. ed.) Scientific Publishers, Jodhpur.

Singh, R.P., Ong. C.K., and Saharan, N. 1989 . Above and below ground interactions in alley cropping in semi-arid India. *Agroforestry Systems* 9: 259-274.

Singh, R.P., Vijayalakshhmi, K, Korwar, G.R. and Osman, M. 1987. Alternate land use system for drylands of India. Central Research Institute for Dryland Agriculture. Research Bulletin No.6 Hyderabad, India, p. 61.

Solanki, K.R. and Ram Newaj 2002. *In–situ* moisture conservation to boost yield of aonla in agri-horticulture system in rainfed conditions. *Indian Farming* September, 2002 : 19-21.

Solanki, K.R., Ram Newaj and Yadava, R.S. 1999. Aonla based agri-silvi-horticultural system- a boon for rainfed farmers. *Indian Farmers Digest* 32 (8) August, 1999 : 17-19.

Sur, H.S. and Sandhu, I.S. 1994. Effect of different grass barriers on runoff, sediment loss and biomass production in foothills of Shiwaliks. *8th ISCO Conference.*, New Delhi, 221p.

Vashishtha, B.B. 1991. Scope and importance of minor fruit in wasteland. *Lecture delivered in summer institute on "Arid horticulture in wasteland development"* N.D. Unversity of Agriculture and Tchnology, Kumarganj, Faizabad.

Venkateswarlu, J. 1993. Problems and prospects in desertification control- Role of Central Arid Zone Research Institute. Pp 249-267, In: Desertification and its control in the, Shara and Sahel Region (sen, A.K. and kar, A. eds.) Jodhpur, India: Scientific Publisher.

Venkateswarlu, J. and Kar, A. 1996. Wind erosion and its control in arid north west India. *Annals of Arid Zone* 35:85.99

Vergera, N. T. 1982. New direction in agroforestry – the potential of tropical tree legumes. East West Centre, environment and Policy Institute, Honululu: p. 52.

Young, A. 1997. *Agroforestry for Soil Conservation.* CAB International, Wallingford (UK); 276 p.

Zimmerman, T. 1986. Agroforestry – a lot of hope for conservation in Haiti. *Agroforestry System,* 4 : 255-268.

11

Choice Experiment Method to Inform Flood Risk Reduction Policies

Ekin Birol, Phoebe Koundouri and Yiannis Kountouris

Introduction

Since the 1990s flood risk and the effects of flooding episodes have reemerged as an important natural hazard concern in central and northern Europe. These concerns have also been exacerbated as a result of widespread and ever increasing awareness of global climate change, and significant wetland loss due to rising sea levels. Global climate change and wetland loss are expected to increase the frequency and extent of floods in the future (Nichols *et al.*, 1999). These floods are expected to cause significant changes in the current land use and population patterns. Contrary to the flooding episodes of the past centuries, recent floods in Europe have milder effects in terms of loss of human life. Economic costs of flooding, however, are rapidly increasing as a result of high costs of damages to infrastructure and production in primary, secondary and tertiary sectors, and disruptions to transport. In Poland and surrounding countries, the estimated costs of the damages of the floods of 1997 and 2001 are in the region of one billion USD for Poland, and 250 million USD, for the surrounding countries (Brakenridge *et al.*, 1997, 2001). As a consequence of the increasing economic and social costs of floods, European governments have taken a more involved approach in flood risk reduction.

Capturing of the welfare effects of flood risk reduction projects and policies is crucial for carrying out the appropriate cost benefit analyses to inform those projects and policies that maximise economic efficiency while minimising flood risks. Even though costs of flood control initiatives are relatively easy to calculate, estimation of the economic benefits of flood risk reduction is a challenging task. This is due to the public good nature of improving flood controls, implying that there are no markets or market prices that could be used for the estimation of the economic benefits that would arise from such projects or policy changes. Non-market valuation techniques, therefore, could be applied in order to estimate the economic benefits of flood risk reduction.

In the existing literature on the valuation of flood risk reduction, a number of non-market valuation techniques have been employed. These include the contingent valuation method, the hedonic pricing method and the aversive behaviour method. Shabman and Stephenson (1996) compare the results of these methods, applied to the valuation of a flood risk reduction project in Roanoke, Virginia, USA. Brouwer and van Ek (2004) and Brouwer *et al.* (2007) employ integrated impact assessment methods to estimate the benefits of flood risk reduction in the Netherlands and Bangladesh respectively, and conduct cost benefit analyses for various flood alleviation projects. Ragkos *et al.* (2006) employ the contingent valuation method to estimate the value of flood control for the Zazari-Cheimaditida Wetland in Greece.

This paper contributes to the literature on valuation of the economic benefits of flood risk reduction by presenting an application of a non-market valuation method, namely the choice experiment method in the Upper Silesia Region of Poland. The paper is structured as follows: In the next section the case study area is described. Section 3 presents the theoretical underpinnings of the Choice Experiment Method. Section 4 describes the survey instrument, and sections 5 and 6 report the results. Section 7 concludes the paper with policy implications for flood risk reduction the Upper Silesia Region of Poland.

The Case Study Area

The case study reported in this chapter presents the results of a choice experiment carried out in the city of Sosnowiec. The first aim of this choice experiment is to estimate the economic benefits that the local residents derive from flood risk reduction in the area. The city of Sosnowiec is located in the Bobrek catchment, in the Upper Silesia Region of Poland. The region is an important industrial center located within the Upper Silesian Coal Basin. There are five rivers (Biala, Brynica, Jaworznik, Wielonka and Rawa) running through the wider area, making the region susceptible to flooding episodes

The main economic activities in the area include heavy industry and mining with some of the world's largest butaminous coalmines located in the region. The mines are concentrated close to the rivers, constantly changing and eroding riverbanks and their morphology. Mining activities have been taking place in this area for over two centuries. Scientific evidence from Central Mining Institute, Silesian University, AGH University of Science and Technology, and Krakow University of Technology claim that mining industry has significantly deformed the local landscape and the riverbed, thereby rendering the region extremely vulnerable to floods even after light rainfalls. Given the size of the local communities, it is estimated that approximately 50000 individuals may suffer the effects of a flood episode.

In 1992 the Polish government facilitated the construction of concrete barriers on the rivers' banks in order to minimize the risk of flooding in the region. Mining industries were deemed responsible for protecting their mines by constructing spoil hips on the rivers' banks. This strategy, however, was not successful since it increased the speed of flowing water, thereby generating negative externalities for downstream communities. Moreover, recreational activities in the catchment became limited as a result of the blocking of the river access by the concrete barriers. Furthermore this policy was not successful in providing flood control as the extensive floods of 1997 and 2002 can attest.

The high economic and social costs of flooding episodes are borne mainly by the local residents, but also by the overall national economy, as well as by the nearby countries. Despite these costs, floods have also brought about some benefits: Unique ecological wetland habitats have been formed on those lands that have been flooded by the rivers. New species of both animals and plants live in these habitats. Ecologists from Silesian University recognise these biodiversity riches and assert that they should be conserved. In addition, these habitats created by the over flown rivers are now of high recreational value, serving as attractive tourism location. A second aim of this choice experiment is therefore to investigate the local residents' valuation of the conservation of the biodiversity found in these habitats and also accessibility to the riverbanks to enjoy recreational activities in the area.

The Choice Experiment Method

The choice experiment method has its theoretical grounding in Lancaster's model of consumer choice (Lancaster, 1966), and its econometric basis in random utility theory (Luce, 1959; McFadden, 1974). Lancaster proposed that consumers derive satisfaction not from goods themselves but from the attributes they provide. To illustrate the basic model behind the choice experiment presented here, consider a household's choice for a river management strategy and assume that utility depends on choices made from a set C, *i.e.*, a choice set, which includes all the possible river management strategy alternatives. The household is assumed to have a utility function of the form:

$$U_{ij} = V(Z_{ij}) + e(Z_{ij}) \qquad (1)$$

where, for any household i, a given level of utility will be associated with any river management strategy alternative j. Utility derived from any of the river management strategy alternatives depends on the attributes of the river management strategy (Z_j), such as the flood risk level, biodiversity level in the habitats and the level of difficulty of access to the river for recreational purposes.

The random utility theory is the theoretical basis for integrating behaviour with economic valuation in the choice experiment method. According to random utility theory, the utility of a choice is comprised of a deterministic component (V) and an error component (e), which is independent of the deterministic part and follows a predetermined distribution. This error component implies that predictions cannot be made with certainty. Choices made between alternatives will be a function of the probability that the utility associated with a particular wetland management option *j* is higher than those for other options. Assuming that the relationship between utility and attributes is linear in the parameters and variables function, and that the error terms are identically and independently distributed with a Weibull distribution, the probability of any particular wetland management plan alternative *j* being chosen can be expressed in terms of a logistic distribution. Equation (1) can be estimated with a conditional logit model (McFadden, 1974; Greene, 1997 pp. 913-914; Maddala, 1999, p. 42), which takes the general form:

$$P_{ij} = \frac{\exp(V(Z_{ij}))}{\sum_{h=1}^{C} \exp(V(Z_{ih}))} \qquad (2)$$

where, the conditional indirect utility function generally estimated is:

$$V_{ij} = \beta_1 Z_1 + \beta_2 Z_2 + \ldots\ldots + \beta_n Z_n \qquad (3)$$

where, n is the number of river management strategy attributes considered, and the vectors of coefficients β_1 to β_n are attached to the vector of attributes (Z).

Survey Design and Administration

The first step in choice experiment design is to define the environmental good to be valued in terms of its attributes and their levels. It is essential to identify the attributes that the public considers important regarding the proposed policy change, as well as those levels that are achievable with and without the proposed policy change. The good to be valued in this choice experiment study is the river management strategy. Following discussions with scientists from the Central Mining Institute, the Silesian University, the AGH University of Science and Technology and the Krakow University of Technology, and drawing on the results of focus group discussions with the local population, three attributes were chosen: surface and underground flooding risk, biodiversity found in the habitats and access to the river. All three of these attributes were specified to have two levels. The payment vehicle was a percentage change in the local taxes paid by the households. Percentage change on the household's present level of tax level was preferred over fixed changes in the tax levels, since the former allows for a continuous monetary variable. Furthermore, higher and lower tax levels than the status quo level were considered in order to understand whether the households are willing to pay to have higher/lower levels of these attributes or willing to accept compensation to let go higher/lower levels of these. Finally, taxation was preferred as a payment vehicle over voluntary donations since households may have the incentive to free-ride with the latter (Whitehead, 2006). Table 11.1 defines the attributes, their levels and the status quo.

A large number of unique river management strategies can be constructed using these attributes and their levels. Using experimental design techniques (Louviere *et al.,* 2000) an orthogonalization procedure was used that resulted in 32 pairwise comparisons of river management strategies. These were randomly blocked into four versions, each containing eight choice sets consisting of two river management

Assuming that the following three river management strategies were the only choices you had, which one would you prefer?

Management Strategy Characteristics	Management Strategy A	Management Strategy B	Neither Management Strategy : Status Quo
Flood risk	Low	Low	High
Biodiversity	Low	High	Low
River access	Difficult	Easy	Difficult
Council tax	5 per cent decrease	5 per cent decrease	Same as now
I prefer(Please tick as appropriate)	Management strategy A*	Management strategy B*	Neither management strategy*

strategies and an opt-out alternative, which represented the status quo. Inclusion of the status quo or another baseline scenario is important for the welfare interpretation

Table 11.1: Attributes, levels and their definitions.

Attribute Name	Definition and Levels
Flood Risk	This attribute refers to the risk of flooding in the area. Levels are
	HIGH: This is the case where no measures are taken and it also reflects the current flood risk level. Danger of flooding is imminent in case of rainfall. No barriers of any kind are built to protect the area from flooding.**LOW**: Both underground and surface barriers are set in place. To avoid past mistakes, the material is proposed to be wood for the surface barriers and concrete for the underground ones. Flooding danger is minimal.
River Access	This attribute refers to public's access to the river for recreational purposes. Levels are:
	EASY: Canalization of the river is very similar to the natural one. Materials such as concrete will not be used. Access to the river's will be possible and easy for everyone.
	DIFFICULT: Rivers will be canalized by forming vertical walls, the same measure that has been used a few years ago. Concrete will be used and it will be impossible for locals to access the river. At the moment access to the river is difficult.
Biodiversity	This attribute refers to the number of different species of plants and animals, their population levels, number of different habitats and their size in the river ecosystem in the next 10 years. The levels are:
	LOW: Due to the present regulation, companies are allowed to create spoil hips from the remnants of their mining activities. This poses a threat to the newly formed habitats, which are being filled with litter. As a result the current biodiversity levels are low and if the current situation prevails, biodiversity will reach a minimum level
	HIGH: As a result of reclamation activities on the existing spoil heaps especially afforestation in the rivers, biodiversity will reach a higher level in 10 years
Local Tax	This is the local, municipal tax paid by every household in the area. The levels are 10 per cent less than the present level, 5 per cent less than the present level, same as the present levels, 5 per cent more than the present level, and 10 per cent more than the present level.

of the estimates and for their consistency with demand theory (Louviere *et al.*, 2000; Bennett and Blamey, 2001; Bateman *et al.*, 2003). Figure 11.1 provides an example of a choice set.

The choice experiment survey started with the enumerators reading a statement identifying the current issues in the area regarding flood risk, biodiversity conservation and use of the river for recreational activities. Subsequently the households were presented with a description of the attributes used in the experiment and were asked to state their preferred river management strategy among three such strategies through eight choice sets. Figure 11.2 presents an example of a choice set.

The choice experiment survey was implemented in March and April 2007 in the city of Sosnowiec, located in the Bobrek catchment, with in house face-to-face interviews. Binding time and budget constraints allowed for a sample of 200 households from the local population. A quota sample was collected and the survey was administered to be representative of the local population in terms of income and geographical distribution (*i.e.*, distance from the river). Those household members who took part in the survey were by and large those who were main household

Table 11.2: Descriptive statistics of respondents and their households, Sample Size=192

Socio-economic Variables	Sample Mean	Population Average
Respondent characteristics		
Age (in years)	46	37.3**
Per cent in full time employment	46.3	
Per cent of female	51.5	51.5*
Per cent with a University degree	26	
Household characteristics		
Household size	2.8	
Distance from the river in meters	462	–
Local tax (in zloty) paid by the household	183.9	
Monthly gross household income (in zloty)	2478.1	1175
Per cent of Households with at least one Child	70.8	
Number of children living in households with children	0.9	
Per cent own a car	64.5	
Per cent visited the wetland	54.6	–
Per cent houses flooded	13	
Per cent flooded households that were compensated by the government, insurance company or mining industry	28	
Number of flood episodes flooded households suffered in the last decade	2.52	
Total damages to the household from floods in the last decade (in zloty)	7115.8	

*: World bank gender Statistics; **: CIA World Factbook.

decision makers and/or heads of the households. In total 96 percent of those approached, *i.e.*, 192 households were interviewed, resulting in1536 choices.

In addition to the choice experiment, the survey also collected social, demographic and economic data, including the respondents' age, gender, education, household income and local tax paid by the household, as well as information on whether the household uses the river for recreational activities and flooding episodes that have effected the household in the past decade. Descriptive statistics for the key variables are presented in Table 11.2.

Results

The data for econometric analysis were coded according to the levels of the attributes. Attributes with two levels (*i.e.*, flood risk, biodiversity level, river access) entered the utility function as binary variables, effects coded as 1 to indicate low level of flood risk, high level of biodiversity and easy river access, and -1 to indicate high level of flood risk, low level of biodiversity and difficult river access (Adamowicz *et al.*, 1994; Louviere *et al.*, 2000). The attribute with five levels (*i.e.*, percentage increase in local tax) was entered in cardinal-linear form, and then multiplied by the households' actual level of local tax, in order to calculate the level of this attribute for each household. Since this choice experiment involves generic instead of labelled options, the alternative specific constants (ASC) were set equal to 1 when either river management strategy A or B was chosen and to 0 when the households chose the status quo (Louviere *et al.*, 2000). A relatively more positive and significant ASC indicates a higher propensity for households to take no action to manage the river.

Retaining the assumption that observable utility function follows a strictly additive form, a conditional logit model for the choice of river management strategy was estimated using LIMDEP 8.0 NLOGIT 3.0. The model was specified so that household choice was only affected by the ASC and the four attributes of the choice experiment. The results of the conditional logit model for the pool of 192 households are reported in first column of Table 11.3.

The results in Table 11.3 indicate that all attributes are highly significant determinants of river management strategy choice for the pooled sample. Furthermore, the estimated coefficients have the expected signs. These indicate that households prefer low flood risk, high biodiversity and easy river access. Consistent with demand theory, the coefficient of the monetary attribute is negative indicating that households choose alternatives with lower tax rates to alternatives with higher tax rates. The positive and significant alternative specific constant captures other factors affecting choice that are not included in the model and can also be interpreted as an indicator of status quo bias.

Successful In order to further examine the behaviour of different groups of households and subsequently to estimate their valuation of each one of the attributes, split sample conditional logit models were estimated for the following four household types: (i) non-flooded in the past ten years, (ii) flooded in the past ten years, (iii) user of the river for recreational purposes, and (iv) non-user. These are reported in columns 2 to 5 of Table 11.3.

Table 11.3: Conditional logit model results for pool, non-flooded, flooded, user and non-user households.

Variable	Pool	Non Flooded	Flooded	User	Non-User
			Coefficient (Standard Error)		
ASC	0.381*** (0.105)	0.344*** (0.105)	0.965*** (0.316)	0.691*** (0.143)	0.095 (0.143)
Flood Risk	0.343*** (0.043)	0.278*** (0.043)	0.862*** (0.131)	0.312*** (0.053)	0.395*** (0.063)
Biodiversity	0.076** (0.04)	0.067* (0.04)	0.009 (0.11)	0.173*** (0.045)	−0.032 (0.058)
River Access	0.137*** (0.042)	0.175*** (0.042)	−0.217** (0.123)	0.216*** (0.052)	0.049 (0.061)
Tax Rate	−0.029*** (0.003)	−0.029*** (0.003)	−0.012 (0.014)	−0.022*** (0.003)	−0.048*** (0.005)
No of observations	1536	1336	200	840	696
Log Likelihood Function	−1498.707	−1319.578	−159.2430	−780.2970	−674.6439
ρ^2	0.112	0.10	0.28	0.154	0.11769

***: 1 per cent significance level; **: 5 per cent significance level, and *: 10 per cent significance level with two-tailed tests.

Swait Louiviere log likelihood ratio test rejects the null hypothesis that the regression parameters for the pooled model and for the flooded and non-flooded subsamples are equal at 0.5 per cent significance level[1]. Hence flooded and non-flooded households have distinct preferences for river management attributes. For those households whose houses have not been flooded all of the river management attributes are significant determinants of river management strategy choice. They prefer those river management strategies, which provide low flood risk, high levels of biodiversity and easy access to the river. The sign on the coefficient on the monetary attribute is negative as expected a priori. Coefficient on the flood risk attribute is the largest in magnitude, implying that this is the most important determinant of choice for the household. This is followed by river access and biodiversity. For those households whose houses were flooded at least once in the past ten years, flood risk reduction and water access are significant determinants of river management strategy choice. These households prefer those river management strategies with low flood risk, however with difficult river access. Their valuation of the biodiversity attribute as well as the coefficient of the monetary attribute are statistically insignificant, the latter possibly due to the small size of this sub sample. Similarly, Swait Louiviere log likelihood ratio test also rejects the null hypothesis that the regression parameters for the pooled model and for the user and non-user sub samples are equal at 0.5 per cent significance level[2]. User and non-user households therefore have distinct preferences

1   Landrace=-2[-1498-(-1319.6+-159.24)]=39.72, which is larger than 16.75, the critical value of chi square distribution at 5 degrees of freedom.

2   Landrace=-2[-1498-(-780.3+-674.6)]=87.54, which is larger than 16.75, the critical value of chi square distribution at 5 degrees of freedom.

for river management attributes. River management strategy choice for those households who use the river for recreational activities is influenced by all of the attributes. These households prefer those river management strategies with low flood risk, easy access and high biodiversity, where the most important attribute is the flood risk, followed by river access and biodiversity. Finally, those households who do not use the river for recreational activities do not derive significant values for biodiversity and river access attributes. Non-user households, however, prefer those river management strategies which provide low flood risk, and which are less costly, in terms on increase in local tax, as expected a priori.

WTP Estimates

The choice experiment method is consistent with utility maximisation and demand theory (Bateman *et al.*, 2003). Welfare measures can be estimated from the parameter estimates reported in Table 3, using the following formula:

$$CS = \frac{\ln \sum_i \exp(V_{i1}) - \ln \sum_i \exp(V_{i0})}{\beta_{tsx}} \tag{4}$$

where, CS is the compensating surplus welfare measure, β_{tax} is the marginal utility of income (represented by the coefficient of the monetary attribute in the choice experiment, which in this case is local tax) and V_{i0} and V_{i1} represent indirect utility functions before and after the change under consideration. For the linear utility index the marginal value of change in a single river management strategy attribute can be represented as a ratio of coefficients, reducing equation (4) to

$$WTP = -1\left(\frac{\beta_{attribute}}{\beta_{tax}}\right) \tag{5}$$

This part-worth (or implicit price) formula represents the marginal rate of substitution between payment and the river management strategy attribute in question, or the marginal welfare measure (*i.e.*, WTP) for a change in any of the attributes. For the binary river management strategy attributes (i.e, flood risk, river access and biodiversity) the marginal implicit price formula becomes (see, Hu *et al.*, 2004):

$$WTP = -2\left(\frac{\beta_{attribute}}{\beta_{tax}}\right) \tag{6}$$

Table 11.4 reports the estimated marginal WTP for each river management strategy attribute for the pool and for the four household types introduced in the previous section.

As revealed by the WTP estimates for the pooled sample, on average households are WTP significant positive amounts for improving all attributes. They are WTP the highest in order to reduce the risk of flooding to a low level, their WTP for easy river access is less than half of their WTP for low flood risk, whereas their WTP for high levels of biodiversity is less than quarter of their WTP for low flood risk. Across the household types, ranking of the attributes, as well as households' valuation of these differ significantly. Flooded households are WTP highest for low flood risk, however their valuation is insignificant. These households are followed by users, non-users and non-flooded households. High biodiversity levels and easy access to the river are valued most highly by users of the river, as expected, they are followed by non-flooded households, whereas flooded and non-user households' WTP are insignificant.

Table 11.4 Marginal WTP for river management scenario attributes for pool, non-flooded, flooded, user and non-user households (zloty/household) and 95 per cent C.I.

Attribute	Pool	Non-Flooded	Flooded	User	Non-User
Flood Risk	23.9***	19.6***	140	28.3***	16.5***
	(20.6–27.4)	(15.9–22.4)	(−16–296)	(22.42–34.1)	(13.5–19.5)
Biodiversity	5.3***	4.8**	0.8	15.7***	−1.3
	(2.8–7.8)	(2.1–7.5)	(−15.82–19.02)	(11.2–20.2)	(−3.8–1.1)
River Access	9.6***	12.1***	−34.2	19.5***	2
	(7.1–12.1)	(9.4–14.7)	(−85.4–16.9)	(15.2–23.9)	(−0.4–4.5)

***: 1 per cent significance level; **: 5 per cent significance level, and *: 10 per cent significance level with two-tailed tests.

Conclusions and Policy Implications

Following the flooding episodes of 1997 and 2004 the Polish authorities embarked on an attempt to reduce flood risk in the Upper Silesia region. The application of the choice experiment method introduced in this chapter focused on the estimation of the benefits that the local population derives from the reductions of flood risk in the area.

The results presented reveal that there are significant welfare improvements from flood risk reduction, which dominate welfare improvements from both improving river accessibility for recreational reasons and conserving high levels of biodiversity. This can be translated as the locals' preferences for use values derived from flood reduction relative to use and non-use values from recreation or biodiversity conservation. Aggregation over the population of Sosnowiec shows that local residents are willing to incur an increase in local taxation of 2693416 zloty per year to reduce flood risk. These results can be relevant for conducting the appropriate cost benefit analysis for flood control infrastructure in the region, as the analysis takes into account both use and non use values derived from policy changes.

Acknowledgements

The authors acknowledge the financial support from the AQUASTRESS integrated project funded by the European Union Framework Programme 6. The authors would also like to thank Anna Adamus, Leszek Trzaski, Nickolaos Syrigos and Haris Giannakidis for their assistance in the development and implementation of the choice experiment survey.

References

Bateman, I.J., Carson, R.T., Day, B., Hanemann, W.M., Hanley, N., Hett, T., Jones-Lee, M., Loomes, G., Mourato, S., Ozdemiroglu, E., Pearce, D.W., Sugden, R., Swanson, S., 2003. Guidelines for the Use of Stated Preference Techniques for the Valuation of Preferences for Non-market Goods, Edward Elgar, Cheltenham.

Bennett, J.J., Blamey, R.K., 2001. The Choice of Modelling Approach to Environmental Valuation, Edward Elgar Publishing Limited, Cheltenham.

Brakenridge, G.R., Anderson, E., Caquard, S., 1997, Flood Archive table 1997, Dartmouth Flood Observatory, Hanover, USA, digital media, http://www.dartmouth.edu/~floods/Archives/1997sum.htm

Brakenridge, G.R., Anderson, E., Caquard, S., 2001, Flood Archive table 2001, Dartmouth Flood Observatory, Hanover, USA, digital media, http://www.dartmouth.edu/~floods/Archives/2001sum.htm

Brouwer, R., van Ek, R. 2004. Integrated ecological, economic and social impact assessment of alternative flood control policies in the Netherlands. *Ecological Economics* 50:1-21.

Brouwer, R., Akter, S., Brander, L., Haque, E. 2007. Socio-economic Vulnerability and Adaptation to Environmental Risk: A Case Study of Climate Change and Flooding in Bangladesh. *Risk Analysis*, 27: 313-326.

Carlsson, F., Frykblom, P., Liljenstolpe, C., 2003. Valuing wetland attributes: an application of choice experiments. *Ecological Economics* 47: 95-103.

Greene, W.H., 1997. Econometric Analysis. Fourth Edition, Prentice Hall.

Hanley, N., Wright, R., Adamowicz, W., 1998. Using Choice Experiments to Value the Environment. *Environmental and Resource Economics*. 11(3-4), 413-428.

Hensher, D., Rose, J., Greene, W., 2005. Applied Choice Analysis: A Primer. Cambridge University Press.

Lancaster, K., 1966. A new approach to consumer theory. *Journal of Political Economics* 74, 217–231.

Louviere, J.J., Hensher, D. Swait, J., Adamowicz, W., 2000. Stated Choice Methods: Analysis and Applications. Cambridge University Press, Cambridge.

Luce, D. 1959. Individual Choice Behaviour, John Wiley, New York, NY.

Maddala, G.S., 1999. Limited Dependent and Qualitative Variables in Econometrics, Cambridge University Press, Cambridge.

McFadden, D., 1974. Conditional logit analysis of qualitative choice behavior. In: P. Zarembka, Editor, Frontiers in Econometrics, Academic Press, New York.

Nicholls, R.,J., Hoozemans F., Marchand, M. 1999. Increasing flood risks and wetland losses due to global sea-level rise: regional and global analyses. *Global Environmental Change* 9: S69-S87.

Whitehead, J.C., 2006. A practitioner's primer on contingent valuation. In: Alberini, A., Kahn, J. (Eds.), Contingent Valuation Handbook. Edward Elgar Publishing, Cheltenham, U.K.

12

Watershed Management: An Integrated Mission

Anil Kumar Gupta, Pallavi Saxena and Apoorva Mathur

Introduction

Watershed management is seen as a way to raise rainfed agricultural production, conserve natural resources, and reduce poverty in the world especially in semi-arid tropical regions. Planning and development of watersheds calls for a rigorous understanding of the occurrence and movement of water in the surface and sub-surface systems along with soil and nutrient losses in a watershed as the need arises for a proper watershed management of that area. In a country like India, where a lot of running water goes waste, it becomes very important to apply the technology of watershed management to solve its annual problems of droughts and floods. About ninety-nine percent of watershed development projects are based on conventional approaches considering only physical aspects without attention to socio-economic or ecological conditions (Farrington and Lobo; 1997). Recently, both the government and non-governmental organizations have realized that protection of watersheds cannot be achieved without the willing participation of local people (Pretty and Ward, 2001). Therefore for successful and sustainable watershed management, people's participation is essential. This paper deals with participatory research on different types of watershed projects in India and also discusses the approaches and methods they follow for the sustainability of the people. Moreover, it also demonstrate that a participatory approach enables the community to visualize and evaluate the impact of innovative technologies.

"A watershed is a topographically delineated area that is drained by a stream system, it is a hydrological unit that has been described and used both as a physical –biological unit, as a socio-economic and socio-political unit for planning and implementing resource management activities (jaiswal, N.K;1983)". Watershed management is seen as a way to raise rainfed agricultural production, conserve natural resources, and reduce poverty in the world's semi-arid tropical regions. These areas, found mainly in South Asia and sub-Saharan Africa, are characterized by low agricultural productivity, severe natural resource degradation, and high levels of poverty (Carney and Farrington; 1998). They were little affected by the Green Revolution that transformed agriculture in more favorable areas. In the semi-arid tropics, watershed management projects aim to capture water during rainy periods for subsequent use in dry periods (Farrington, Turton, and James, 1999). These projects, which often operate at the level of a microwatershed within a single village, focus on conserving soil moisture for rainfed agriculture, recharging aquifers to augment groundwater irrigation, and capturing surface runoff water in small ponds (Johnson *et al.*, 2003). Where water harvesting is the main objective, the projects involve construction of small check dams in drainage lines. To be sustainable, water harvesting requires protecting the upper reaches against erosion that would deposit silt behind the structures, reducing their water-holding capacity. Accordingly, productivity and conservation objectives are highly complementary. Integrated watershed development projects are being implemented in the country, in the plains and hills to improve the productivity potential of the project area in states evolving watershed treatment technology through community participatory approaches (Swallow *et al.*, 2001). Although watershed development (the development of watershed resources in a watershed) and development of watershed resources management are not the same, there is a great deal of overlap between the both, physically and conceptually (Murty, J.V.S; 1991). The modern concept of watershed management has grown out of these two approaches:

☆ Managing water resources

☆ Managing socio-economic systems

Watershed management had a strong hydrologic focus particularly on the use of structural as well as structural practices to control the quality and quantity and timing of water flows (ASCE, 1975). Therefore watershed management is more than just the proper economic analysis of water resource projects and understanding of interaction between land, water and people for production of goods and services is equally important part of this topic (Ealy, C.D *et al.*, 1975).

Watershed Characteristics

Each watershed shows distinct characteristics which are so much variable that no two watersheds are identical (Sen. and Purandare, 1984). All the characteristics affect the disposal of water (Kathuria, 1978). These are:

I. Size

Size of the watershed forms the basis for further classification into different categories (Bali, 1980):

☆ Sub watershed (100-500 sq. km)

☆ Milli watershed (10-100 sq. km)

☆ Micro watershed (1-10 sq. km)

☆ Mini watershed (less than 1 sq. km)

The size helps in computing parameters like precipitation received, retained and drained off. Larger the watershed more the heterogeneity of other characteristics.

II. Shape

Watershed differ in shape based on morph metric parameters *e.g.* Geology and structure. General shapes are pear, elongated, circular. The shape affects the runoff characteristics (ICRSAT. 1981)

III. Physiographic

Type of land, its altitude and physical deposition immensely speak about watershed for planning the activities in greening.

IV. Slope

It controls the rainfall, distribution and movement, land utilization. The degree of slope affects the velocity of overland flow and runoff, infiltration rate and thus soil transportation (Kampen, J. *et al.,* 1974).

V. Climate

Meteorological parameters decide a quantative approach for arriving at water availability in watershed. Climate is the determining factor.

VI. Drainage

The order pattern and density of drainage have profound influence on watershed. it determines flow characteristics and thus erosional behavior (Kowal, J. 1970).

VII. Land Use

It is vital for planning, programming and implementing projects on watershed. It portrays man's impact on the specific watershed and forms a basis for categorizing the land for formulation of pragmatically action plan (Murty, J.V.S1989).

VIII. Vegetation

Information on vegetation helps in choosing type, mode and manner of greening the watershed (Jodha, N.S; 1989).

IX. Geology and Soil

Rocks and their structure control the formation of watershed because their nature determines size, shape, physiographic and ground water conditions (ICRSAT. 1984).

X. Hydrology

Hydrological parameters help in quantification of water available, utilized and additional exploitable resources for greening the area (Michelson, R.H; 1966). Helps to determine location and design of conservation of resources (Kathuria, K.C. 1978).

XI. Socio-Economics

Statistics on people and their health and hygiene, wants and wishes, cattle and farming practices and share of participation are equally important in managing watershed.

Selection of Watersheds

A watershed is a geohydrological unit or area that drains at a common point. In villages selected for intervention, watershed totaling approximately 1000-2500 hectares shall be identified and selected by voluntary organization (VO) in consultations with the villagers from the watershed area. The following criteria may be used in the selection of watersheds (ministry of agriculture. 1988):

1. Ones which have acute shortage of water, especially drinking water.
2. Watershed which have large population of schedule caste and scheduled tribes dependent on it.
3. Critical watersheds that have undergone heavy soil erosions and have preponderance of waste lands and highly degraded lands.
4. Those which have preponderance of common lands.
5. Were actual wages are significantly lower than the minimum wages.
6. Those contagious to another watershed which has already been developed/ may be selected for conservation.
7. Watersheds which have been previously taken up for comprehensive development/treatment works. However if the specific area of the watershed now identified had not previously benefited from any developmental works, even though it was a part of larger watershed taken up under any other earlier programs, it may be selected for a project now.
8. 500 hectares is a general norm and if on actual survey, watershed is found to have more or less area, it may be taken up for conservation, keeping in view that the total area to be developed by a project implementation agency is 1000-2500 ha.
9. Where public participation and commitment individually and collectively is available to carry out the percepts and objectives of this scheme.

Watershed – An Integrated Multidisciplinary Approach

Watershed approach invites an integrated input of various disciplines for the development of different minor watersheds in accordance with their characteristic (Murty, J.V.S; 1988). The approach may be the key to (Lee and Chung; 2007a):

☆ Protect natural resources

☆ Attain good yield

☆ Coordinates man power with limited funds

☆ Community participation

The approach trends with a view to achieve holistic management of resources for maximum productivity while preserving environment with funds available. In the Indian context the approach may be programmed into 3 phases (Lee and Chung; 2007b):

Phase I

☆ Rapid identification and classification of geographical area and preparation of priority program

☆ Preparation of master plan, applying remote sensing

☆ Revival of basic scientific practices *e.g.* compost usage

☆ Introducing culture of accessibility of any watershed data available with govt. to the common man

☆ Bringing conceptual awareness through mass media, training in appropriate technology

☆ Creation of central data bank

Phase II

☆ Delivery of proper rural technological systems

☆ Watershed management of upper reached commencing from SAT and wastelands

☆ Extension of activities to the lower reaches, coastal tracts

☆ Creation of regional data banks

☆ Creation of efficient agro industrial infrastructure *e.g.* power supply, hybrid seeds

☆ R&D on watershed conceptual aspects

☆ Simultaneous formulation and enforcement of laws on watershed concept

Phase III

☆ Distribution of watershed to the local people for development

☆ Establishing technological clinics in every watershed

☆ Switching on to watershed based political boundaries

☆ Hastening of national grid of waterways e.t.c

☆ Formulation of natural resources ministry for centralizing involved aspects under proper umbrella

☆ Priority for appropriate technology, applied research and healthy environment.

Watershed Developmental Plan

Watershed plan will specifically include the following sections (Boulanger and Brechet; 2005):

☆ Statement of aims.

☆ Result of resources appraisal exercises to identify needs, availability and gaps.

☆ Present lands and water use patterns, productivity level of different types of lands, across seasons and with map displays.

☆ Proposed land and water use patterns estimated yield levels; with appropriate maps.

☆ Land treatment works, water harvesting and conservation.

☆ Budget; cost benefit assessment.

☆ Project management systems; including appropriate community organization training programs.

☆ Out put utilization strategies; benefit sharing principles and agreements.

☆ Work schedules and monitoring processes.

Importance of Watershed Development

Showing in Figure 12.1.

Watershed Management in India

In Urban India today, it has become a symbol of culture and refinement to talk about and to support environmental causes. But not so in the rural areas where farmers are trying desperately to make both ends meet. Environmental problems in urban areas have received much attention and action while the rural areas, home to 70 per cent of the National population continue to deteriorate. Victims of the whimsical monsoons and fickle market prices, these poor farmers have very little control over their destiny. Furthermore, due to increasing pressure of population, there is demand for more land for agricultural and non-agricultural use. Unhealthy practices on available land have resulted in creation of vast stretches of wastelands due to soil salinity, water logging, and desertification and soil erosion. In fact, according to the Ninth Five Year Plan Document, soil erosion is contributing to degradation in about 45 per cent of the cultivable area of the country. The estimates of wastelands range from 76 million hectares to 175 million hectares. In a densely populated country like India, one cannot afford to let so much land remain idle. To make this land cultivable, the productive approach is through watershed development.

The watershed programmes are implemented by the Zilla Panchayat through watershed associations. A Project Implementing Agency (PIA), which may be a Government Department or an NGO, is assigned about 10 micro watersheds, each micro watershed covering about 500 hectares. The PIA forms a watershed development team that interacts with the watershed associations and provides technical assistance

DEFORESTATION
Population explosion
stress on the resources
over exploitation
over grazing
organic pollution
burning organic matter

↓

Eroded lands
Depleted resources
Increased desertification
Polluted air, water and land
Released deleterious gases

↓

GLOBAL HOLOCAUST
Ozone holing
Global warming
Change in climate
Environmental degradation

↓

Environmental revival
Essential for survival

↓

GREEN COVER
Provides food and other needs
Control erosion
Consumes gases
Releases climate
Restores climate

↓

Green grows with moisture
Moisture source is rainwater
Water moves from ridge to river controlled by
watershed

CLOUDS
Rainwater
Evapotranspiration (30 %)
Recharge(30 %
Runoff + base flow

↓

Over land flows
Flood flows
Base flows

↓

Part can be utilized by dam

↓

Balance wastes into seas

↓

Salvaging essential

↓

METHODOLGY
Contour techniques
Check dams
Construction
Rainwater harvesting
Growing green
Pasture
Silvipasture
Crops
Horticulture
Social forestry
Agroforestry
afforestation

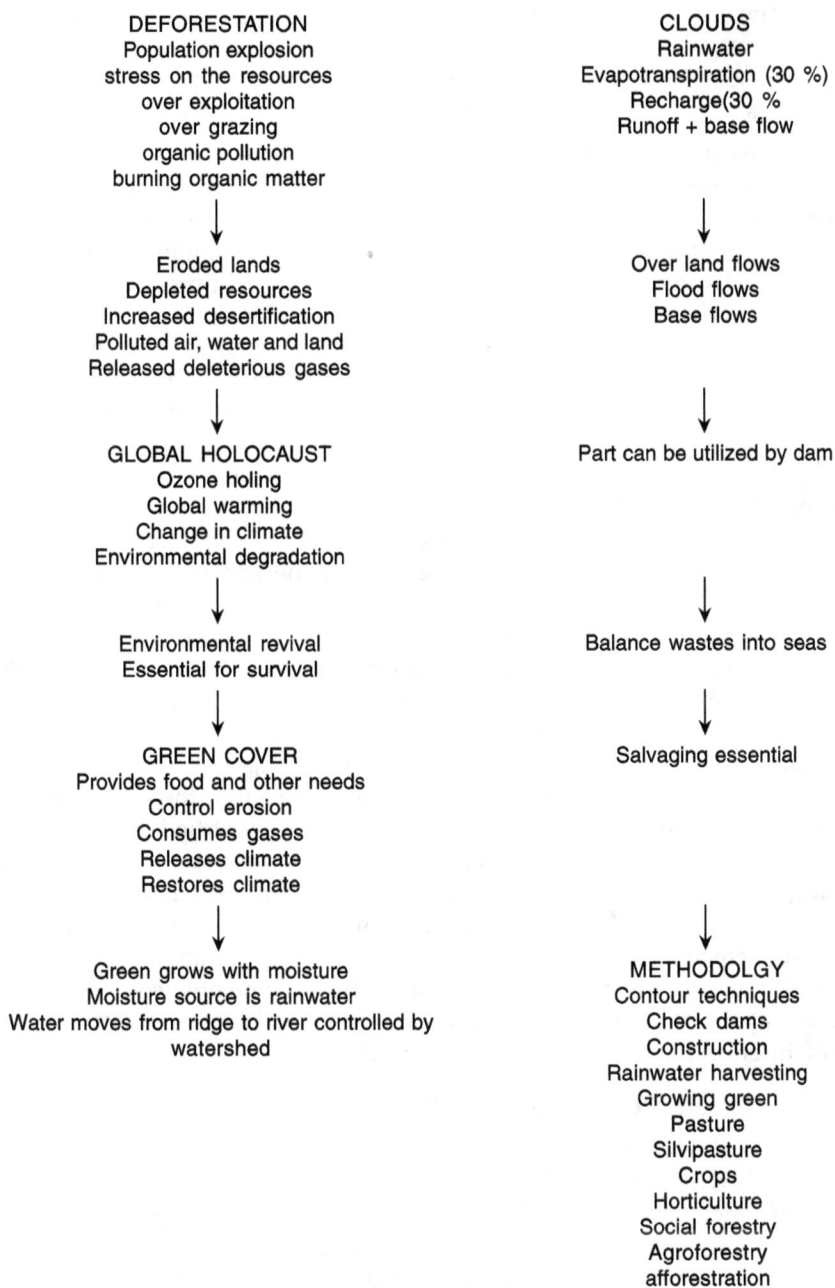

Figure 12.1: Importance of watershed development.

to the watershed association in the planning and implementation of the watershed programme. The residents of the area covered by the watersheds are also organised into self-help groups and user groups. In fact, these user groups are the beginning point as well as the end point for Watershed Development programmes. Their initiative is crucial to the success of the programme and they are the ultimate beneficiaries.

In the last two decades, watershed management has gained the top most priority in water resources sector. Implementation of any water management measure requires a suitable hydrological unit. A properly delineated watershed forms a convenient hydrological unit for computation of water balance parameters and thus implementation of water management schemes. The watershed approach has become a pre requisite for any developmental programme, because land and water resources have maximum interaction and synergic effect, when developed on watershed basis. Watershed approach is, therefore, increasingly applied in various development programmes like command area development, soil and water conservation, flood control, soil erosion control, river valley projects, land reclamation, people and resource dynamics etc. It is also equally important for various hydro-power and irrigation projects, assessment of ground water resource, pollution and artificial recharge studies. For proper planning and execution of any development programme on watershed basis, it is essential to have watersheds in the form of maps along with relevant attributes. There is also improvement in the increasing application of GIS techniques in water sector requires digital maps and geodatabase. Under the Hydrology Project Phase – I, Central Ground Water Board (CGWB) had the mandate to prepare the GIS data set on 1:250, 000 scale for various thematic layers relevant to groundwater for integration in the dedicated software and consequently in the Hydrological Information System (HIS) developed in the project. Watershed was one of the layers envisaged for integration in the software to facilitate the Ground water assessment on watershed basis (CGWB; 2008). CGWB was identified by Ministry of Defence as one of the nodal agency for creation of digital data under the Hydrology Project. Keeping in view the growing requirements of watershed delineation on 1:250,000 scale and its utilities in various departments for implementing diverse developmental activities an urgent need was felt to delineate the watershed boundary on 1: 250,000 and bring the data in GIS platform which can be utilized by different departments/agencies for various purpose. The present Watershed Atlas is the outcome of the effort made by CGWB in this direction.

Recent Watershed Developmental Projects

Recent watershed developmental projects have been taken up under different programs launched by govt. of India. The drought prone areas program (DAP) and the desert developmental program (DDP) adopted the watershed approach in 1987. The Integrated wasteland developmental project scheme (IWDP) taken up by the integrated wasteland developmental project board in 1989 also aimed at developing wasteland on watershed basis. This program has been brought under the administrative jurisdiction of the department of wasteland development. The fourth major program based on the watershed concept is the national watershed development program in rainfed areas (NWDPA) under the ministry of agriculture. Water shed conservation program will be implemented by CAPART through assistance to voluntary organizations over a total area of 1000-2500 hectares

References

ASCE, 1975. Watershed management Utah State Univ. *Proc. Vol. Symp.* p 42.

Bali, 1980 Y.P. 1978. Watershed management- concept and strategy. Central soil and Wat. Cons. Res. and Trg. Instt. Dehradhun. Mimeo.

Boulanger, P.M., Brechet, T., 2005. Models for policy-making in sustainable development: The state of the art and perspectives for research. *Ecological Economics* 55 (3), pp. 337–350.

Carney, D., Farrington, J., 1998. Natural Resources Management and Institutional Change. Routledge, London.

Central Ground Water Board (CGWB) Report; 2008.

Drechsler, M., Wa¨tzold, F., 2006. Ecological–economic modelling for the sustainable use and conservation of biodiversity. *Ecological Economics* 62 (2), pp. 203–206.

Ealy, C.D *et al.,* 1975.resource identification studies on urban watershed using Anacostia river basin as and example. Prog. Rept. NSG -5017.

Farrington, J., Lobo, C., 1997. Scaling up participatory watershed development in India: lessons from the Indo-German watershed development programme. In: *Natural Resource Perspectives,* vol. 17.Overseas Development Institute, London.

ICRSAT. 1981 improving the management of India's deep black soil. Hyd. Ind.

ICRSAT. 1984. Watershed based dry land farming in black and red soil of peninsular India. *Workshp. Proc.* Vol. 25-70

Jaiswal, N.K; 1983, Planning and management of Watershed in drought prone areas. NIRD. Workshop. Ind. Proc. Vol. pp. 1-47.

Jodha, N.S; 1989. Some dimension of traditional farming systems in semi arid tropical India. Worksp. Socioecon. Constraints. SAT Ag. ICRSAT, Hyd. Ind pp. 19-23.

Johnson, N., Lilja, N., Ashby, J., 2003. Measuring the impact of user participation in agricultural and natural resource management research. *Agricultural Systems* 78, pp 287–306.

Kampen, J. *et al.,* 1974. Soil and water conservation and management in farming system research for semiarid tropics. *Intl. Workshp. Farmg. Syst. ICRSAT,* Hyd. Ind pp. 18-21

Kathuria, 1978. Watershed planning for optimum utilization of water. Kurkshetra. 26(21) pp 18-20

Kathuria, K.C. 1978 Watershed planning for optimum utilization of water. Kurukshetra. 26(21) pp. 18-20.

Kowal, J. 1970. The hydrology of small catchments basin. Nigeria. *Ag. Jour.* III. 7 pp. 120-133.

18. Lee, K.S., Chung, E.S., 2007a. Development of integrated watershed management schemes for intensively urbanized region in Korea. *Journal of Hydro Environmental Research* 1 (2), pp. 95–109.

19. Lee, K.S., Chung, E.S., 2007b. Hydrological effects of climate change, groundwater withdrawal, and landuse in the small Korea watershed. *Hydrological Processes* 21 (22), pp. 3046–3056.

20. Michelson, R.H; 1966. Level Pan System for spreading and storing watershed runoff. *Soil Sc. Soc. Am. Proc.* 30 pp. 388-392

21. Ministry of agriculture. 1988. Soil and land use survey for watershed management. Nat. Land Use and Constn. Board. T.S. Land Res. 1 and 2 p.188.

22. Murty, J.V.S; 1991. Watershed management in tribal areas. Lect. Trg. Cours. Watershed management. CRIDA, Hyd.

23. Murty, J.V.S 1989. Entrepreneurship development in watershed management in India. Keynote Address. Worksp. Entrepreneurship Dev. Watershed Managt. Proc. Vol pp 3-11. Murty, J.V.S; 1988. Watershed management in India. Nat. Workshp. TDS Dev. NIWLRD. ND. *Ind .Proc.* Vol. pp. 5-10.

24. Pretty, J., Ward, H., 2001. Social capital and the environment. World Development 29 (2), pp. 209–227.

25. Sen. and Purandare, 1984. Watershed management- a viable proposition to develop northern region. Worksp. Watershed management shillong. Ind. NIRD Pub.

26. Swallow, B., Garrity, D.P., Noordwijk, M., 2001. The effect of scales, flows and filters on property rights and collective action in watershed management. Water Policy 3, pp. 457–474.

13

Hazardous Waste: Treatment and Disposal Technique

S. Pandey, F.Z. Siddiqui and Anil K. Gupta

Introduction

Hazardous Waste (HW) generated by the industries can cause environmental pollution and adverse health effects if not handled and managed properly. In order to manage HW, the Ministry of Environment and Forests, Government of India, notified the Hazardous Waste (Management and Handling) Rules on July 28, 1989, under the provisions of the Environment (Protection) Act, 1986, which was further amended in the year 2000 and 2003, and reissued in 2008 as the Hazardous Waste (Managemnt, Handling and Transboundary Movement) Rules.

The Hazardous Wastes (Management, Handling and Transboundry Movement) Rules, 2008 were notified for effective management of HW, mainly solids, semi-solids and other industrial wastes, which do not come under the purview of the Water (Prevention and Control of Pollution) Act, 1974 and the Air (Prevention and Control of Pollution) Act 1981 and also to enable the Authorities to control storage, transportation, treatment and disposal of waste in an environmentally sound manner. The Hazardous Waste (Management, Handling and Transboundary Movement) Rules, 2008" were further amended in the year 2009.

The objective for introduction of such Rules is to ensure safe management of hazardous waste, generated from different industrial sources. The Rules define various categories of hazardous waste, based on the process listing (waste streams) and concentration of hazard components. The regulatory mechanism for enforcement of the Rules is the responsibility of the State Pollution Control Boards.

There are about 36,165 number of hazardous waste generating industries in India, generating about 6.2 million metric tons of hazardous wastes every year. The quantum of hazardous waste, which has to go for final disposal in secured land fill (SLF) is about 2.7 million metric tons (*i.e.* 44.30 per cent), disposal by incineration is about 0.4 million metric tons (*i.e.* 6.60 per cent) and recyclable waste is about 3.0 million metric tons (*i.e.* 49.10 per cent) of total hazardous waste generation in the Country (6).

Definition of Hazardous Waste

Hazardous waste has been defined in Rule 3 of the Hazardous Wastes (Management, Handling and Transboundry Movement) Rules, 2008 came into force with effect from Sep. 24, 2008, as any waste, which by reason of any of its physical, chemical, reactive, toxic, flammable, explosive or corrosive characteristics causes danger or is likely to cause danger to health or environment, whether alone or when in contact with other wastes or substances, and shall include:

- ☆ Wastes listed in Column 3 of Schedule-1;
- ☆ Wastes having constituents listed in Schedule-2, if their concentration is equal to or more than the limit indicated in the said schedule; and
- ☆ Wastes listed in List 'A', and 'B' of Schedule-3 (Part-A) applicable only in case(s) of import with prior informed consent and for import and export not requiring prior informed consent.

Salient Features of Rules and Guidelines Applicable for Development and Operation of the Common Hazardous Waste Treatment, Storage and Disposal Facilities

The Salient features of the Rules, standards and guidelines relevant to the management of hazardous wastes and the TSDFs are given below.

Environmental Impact Assessment Notification S.O.1533 (E) dated 14 September 2006

According to the environmental impact assessment (EIA) notification dated 14 September 2006, establishment of an integrated facility having incineration and landfill or incineration alone requires 'Environmental Clearance' from the Ministry of Environment and Forests (MoEF) as per the procedures stipulated under these Rules. In case of establishment of a secured landfill (SLF) alone, environmental clearance is to be obtained from the State Environmental Appraisal Committee constituted by the State Government/UT administration.

Hazardous Waste (Management, Handling and Transboundary Movement) Rules, 2008:

The Rules relevant to the TSDFs are summarized below:

According to Rule 3 (j) of the Hazardous Waste (Management, Handling and Transboundary Movement) Rules, 2008 and amendments made thereof, 'facility' means any establishment wherein the processes incidental to the handling, collection, reception, treatment, storage, recycling, recovery, reuse and disposal of hazardous wastes are carried out.

As per Rule 3 (m), 'hazardous waste site' means a place of collection, reception, treatment, storage of hazardous wastes and its disposal to the environment which is approved by the competent authority.

As per Rule 3 (r), 'operator of a facility' means a person who owns or operates a facility for collection, reception, treatment, storage or disposal of hazardous wastes.

The occupier of hazardous waste is required to perform the responsibilities as stipulated under Rule 4 of the said Rules for handling, treatment, storage, transport and disposal of hazardous waste.

The Occupier or Operator of a Hazardous Waste Treatment, Storage and Disposal Facility (TSDF) is required to obtain authorization as per provisions laid down under

Table 13.1: Analysis requirement for hazardous waste treatment storage and disposal facilities.

Parameters for Fingerprint Analysis by the Operators of TSD Facilities	Method of Analysis
Physical Analysis	
Physical State of the waste (liquid/slurry/sludge/semi-solid/solid: inorganic/organic/metallic)	
Identification of different phases of the wastes (in cases of solid wastes contained in aqueous/non-aqueous liquids/solutions for slurries and sludge)	
Colour and Textures	
Specific Gravity	
Viscosity in case of liquid waste	
Flash Point and 1020	USEPA, SW-846; Method 1010
Loss on drying at 105°C in case of solids	
Loss on ignition at 550°C	
Calorific Value in case loss on ignition $\geq$ 20 per cent	
Paint Filter Liquid Test (PFLT) for liquids	USEPA, SW-846; Method 9095
Liquid Release Test (LRT) for liquids	USEPA, SW-846; Method 9096
Chemical Analysis	
pH	USEPA, SW-846; Method 9040, 9041 and 9045
Reactive Cyanide (ppm)	USEPA, SW-846; Vol. 1C Part II; Test Method todetermine HCN releasedfrom Wastes
Reactive Sulfide (ppm)	USEPA, SW-846; Vol. 1C Part II; Test Method to determine H_2S released from Wastes

Table 13.2: Comprehensive analysis requirement for hazardous waste-generator/TSDF operator.

Comprehensive Analysis to be Submitted by the Generators of Hazardous Wastes	Method of Analysis
Physical Analysis	
Physical State of the waste (liquid/slurry/sludge/Semi-solid/solid: inorganic, organic, metallic)	
Description of different phases of the wastes (in cases of solid wastes slurries and sludge) contained in aqueous/non-aqueous liquids/solutions	
Colour and Texture	
Specific Gravity	
Viscosity in case of liquids	
Calorific Value in case of organic wastes	
Flash Point	USEPA, SW-846; Method 1010 and 1020
Per cent Moisture content (loss on drying at 105C)	
Per cent Organic content (loss on ignition at 550 C)	
Paint Filter Liquid Test (PFLT)	USEPA, SW-846; Method 9095
Chemical Analysis	
pH	USEPA, SW-846; Methods 9040, 9041 and 9045
Inorganic Parameters Analysis	
Cyanide (ppm)	USEPA; SW-846; Vol. 1C Part II; Test Method todetermine HCN releasedfrom Wastes
Sulfide (ppm)	USEPA; SW-846; Vol. 1C Part II; Test Method to determine H_2S released from wastes
Sulphur (elemental)	USEPA; SW-846; 9010, 9011, 9012
Concentration of relevant inorganic [as per Schedule 2 of HW (M, H and TM) Rules, 2008 and amendments made thereof].	USEPA; SW-846; Vol. 1A, 1B, 1C and Vol. 2
Organic Parameters Analysis	
Oil and Grease Extractable Organic (in special cases only) Per cent Carbon Per cent Nitrogen Per cent Sulphur Per cent Hydrogen	
Compatibility tests	
Concentration of relevant individual organics [as per Schedule 2 of HW (M, H and TM) Rules, 2008 and amendments made thereof]	USEPA; SW-846; Vol. 1A, 1B, 1C and Vol. 2
Toxicity Characteristics Leaching Procedure (For the listed parameters relevant to the process as presented in Method 1311 of SW 846; USEPA) for landfillable wastes	USEPA; SW-846; Method 1311, 1330

Rule 5 of the said Rules for the purpose of generation, processing, treatment, package, storage, transportation, use, collection, destruction, conversion, offering for sale, transfer of the hazardous waste.

Rule 7 stipulates that the operator of facilities may store the hazardous wastes for a period not exceeding ninety days and shall maintain a record of sale, transfer, storage and re-processing of such wastes and make these records available for inspection to the regulatory authorities. However, Rules empower State Pollution Control Boards (SPCBs) and Pollution Control Committees (PCCs) to extend the storage period maximum up to six months of their annual capacity.

Table 13.3: Criteria for direct disposal of hazardous waste into secured landfill.

Leachate Quality	Concentration
pH	4-12
Total Phenols	<100 mg./l.
Arsenic	<1 mg./l.
Lead	<2 mg./l.
Cadmium	<0.2 mg/l.
Chromium-VI	<0.5 mg./l.
Copper	<10 mg./l.
Nickel	<3 mg./l.
Mercury	<0.1 mg./l.
Zinc	<10 mg./l.
Fluoride	<50 mg./l.
Ammonia	<1,000 mg./l.
Cyanide	<2 mg./l
Nitrate	<30 mg./l
Adsorbable organic bound Chlorine	<3 mg./l
Water soluble compounds except salts	<10 per cent
Strength	
Transversal Strength (Vane Testing)	>25 KN/m2
Unconfined Compression Test	>50 KN/m2
Axial Deformation	<20 per cent
Degree of Mineralization or Content of Organic Mater	ials (original sample)
Annealing loss of the dry residue at 5500 C	<20 Wt. per cent (for non- biodegradable waste)
	<5 Wt. per cent (for biodegradablwaste)
Extractable Lipophylic contents (Oil and Grease)	<4 Wt. per cent

1) leachate quality is based on water leachate test (i.e Leachability tests are conducted by preparing a suspension of waste and water (i.e taking 100 gm of waste and filling up to 1 liter with distilled water), stirring or shaking for 24 hrs, filtering the solids and analyzing the filtrate)

2) Calorific value of the landfillable hazardous wastes should be less than 2500 K. Cal/Kg

Rule 18 (1) of the HW (M,H and TM) Rules, 2008 deals with the joint responsibilities of the State Government, operator of a facility, occupier, or any association of occupiers for identifying sites for establishing the TSDFs.

Rule 18 (2) deals with the design and setting up of TSDFs as per the guidelines of CPCB and obtaining of approval from SPCBs with regard to the design and layout of the TSDFs. Rule 18 (3) deals with the monitoring by SPCB w.r.t the setting up and operation of TSDFs. Rule 18 (4) and 18 (5) deal with the responsibility of the operator of a TSDFs for safe operation of the TSDFs during its operational and ensuring its safety during post-closure period and maintaining of records w.r.t the hazardous waste handled.

Rule 19 deals with the requirement of proper labeling and packaging of hazardous wastes for its safe handling, storage and transportation.

Rule 20 deals with the transportation of the hazardous waste in accordance with the HW (M, H and TM) Rules as well as rules framed under the Motor Vehicle Act, 1988 and other guidelines issued from time to time. These Rules also deals with the requirement of obtaining of 'No Objection Certificate' for final disposal to a facility existing in a State other than the State where the hazardous waste is generated.

Rule 21 deals with the requirement of six colored manifest copies as per Form 13 of the HW (M, H and TM) Rules so as to ensure the wastes are collected, transported, stored, treated and disposed of in an environmentally sound manner.

Rule 25 deals with the liabilities of the operator of a facility in case of damages caused to the environment due to the improper handling, storage and disposal of hazardous wastes.

Gaseous Emission Norms for Common Hazardous Waste Incinerators Notified under the Environment (Protection) Act, 1986 as Environment (Protection) Fifth Amendment Rules, 2008 dated 26 June 2008

Common Hazardous Waste Incinerators are required to comply with the gaseous emission norms notified under the Environment (Protection) Fifth Amendment Rules, 2008, dated 26 June 2008.

Salient Features of the Guidelines

Criteria for hazardous waste landfills: These guidelines provide mainly criteria for location, site selection, site investigation, planning and design, requirements of landfill liner and cover, construction and operation, inspection, monitoring and record keeping, apart from requirement of post-closure, financial assurance as well as contingency plans for emergencies.

These guidelines also emphasize adoption of single liner system or double liner system depending upon the rainfall, type of sub-soil and the water table beneath the base of the landfill. In a place where rainfall is high and/or sub-soil is highly permeable (*e.g.* gravel, sand, silty sand) and/or the water table is within 2.0 m to 6.0 m, the

	1	2	3	4	5	6	7	8	9	10	11	12
1 Oxidizing Mineral Acids	1											
2 Caustics	H	2										
3 Aromatic Hydrocarbons	HF		3									
4 Halogenated Organics	HF	HGF		4								
5 Metals	GF H F			HF	5							
6 Toxic Metals	S	S				6						
7 Sat Aliphatic Hydro-carbons	HF						7					
8 Phenols and Creosols	HF							8				
9 Strong Oxidizing Agents		H	HF		HF	H			9			
10 Strong Reducing Agents	HF GT			HGT				GFH	HF	10		
11 Water and Mixtures containing water	H			HE		S				GFGT	11	
12 Water reactive Substances			Extremely reactive, do not mix with any chemical or waste material									12

Figure 13.1: Compatibility of selected hazardous waste.
E: Explosive; F: Fire; GF: Flammable Gas; GT: Toxic Gas; H: Heat Generation; S: Solubilisation of Toxins.

guidelines suggest to adopt double composite liner. The specifications of the single composite liner, double composite liner system and cover system have been provided:

Guidelines for Proper Functioning and Up-Keep of Disposal Sites

These guidelines suggest responsibilities of the Occupier, transporter, operator of a facility, Toxicity Characteristics Leaching Procedures (TCLP) limits, waste acceptance for direct disposal (*i.e.* finger print and comprehensive analysis parameters) (Tables 13.1 and 13.2 respectively), criteria for direct disposal of hazardous waste into secured landfill (Table 13.3), leachate standards (Table 13.4).

Table 13.4: Proposed leachate disposal standards in addition to the general standards for discharge of environmental pollutants.

S.No	Parameter	Standards (in mg/l)			
		Inland Surface	STP	CETP	Marine Coastal Areas
1.	Adsorbable Organic Halogens (AOX)	0.50	–	–	0.50
2.	Poly Aromatic Hydrocarbons (PAH) each	0.059	–	–	0.059
3.	Benzene	0.14	–	–	0.14
4.	Toluene	0.08	–	–	0.08
5.	Xylene (Sum of o,m,p-xylene)	0.32	–	–	0.32

1. In addition to the above, General Standards for discharge of environmental pollutants Part-A: Effluents notified, vide G.S. R. 422 (E), dated 19.5.1993 and published in the Gazette No. 174, dated 19.5.1993 under the Environment (Protection) Act, 1986, and rules made thereunder, shall also be applicable for disposal of leachate into sewage treatment plant, common effluent treatment plant, Inland surface water bodies or coastal areas.

2. For each Common Effluent Treatment Plant (CETP) and its constituent units, the SPCB/PCC shall prescribe standards as per the local needs and conditions; these can be more stringent than those prescribed above. However, in case of clusters of units, the SPCB/PCC may prescribe suitable limits.

3. The Bioassay test shall be substituted by 'Fish Toxicity' test, and a dilution factor of 2 (two) may be considered.

Guidelines for Storage of Incinerable Hazardous Waste

These guidelines emphasize requirement of adequate storage capacity, minimum of 15 m distance between storage sheds, fire break of at least 04 m between two blocks of stacked drums, maximum of 300 metric tons incinerable waste storage limit in a block of drums, at least 1m clear space between two adjacent rows of drums in a pair for routine inspection purposes, requirement of type of drums to be used for storage of incinerable wastes, spillage or leakage control measures to be adopted in the event of any leakages or spillages, record keeping and maintenance, requirement of fire detection, protection and safety measures as well as performing safety audits every year by the operator of a facility and externally once in two years by a reputed expert agency.

Sampling of Hazardous Waste

This section addresses the sampling plan for different kinds of hazardous wastes as defined in the regulations for the identification and listing of hazardous wastes to include solid, semisolid, liquid, and other materials. Analysis and characterisation of hazardous wastes involves collection of representative samples of wastes for the estimation of analytical properties.

Sampling Procedure

Due to the diversity of the wastes and the waste storage and management scenarios, the sampling procedures and techniques to be employed vary. Proper procedures and considerations for sample collection and preservation have been taken from USEPA prescribed methods from SW-846.

Sampling Strategies

The development and application of a sampling strategy is a prerequisite to obtaining a representative sample capable of producing scientifically viable data. These strategies should be selected or prepared prior to actual sampling to organize and coordinate sampling activities, to maximize data accuracy, and to minimize errors attributable to incorrectly selected sampling procedures. At a minimum, a sampling strategy should address the following

- ☆ Objectives of collecting the samples
- ☆ Sampling approach (*e.g.* Authoritative or Random)
- ☆ Types of samples needed (*e.g.* grab or composite)
- ☆ Selection of sampling locations
- ☆ Number of samples
- ☆ Sampling frequency
- ☆ Sample collection and handling techniques to be used
- ☆ Physical and Chemical Properties of the wastes
- ☆ Special circumstances or considerations (complex multi-phasic waste streams, highly corrosive liquids, *etc.*)

Planning for Sampling

Hazardous wastes are complex and heterogeneous with a variety of physical and chemical properties. The waste samples are usually collected from tanks, drums, ponds, piles or form various processing or transporting equipment such as conveyor belts. It is important that representative sampling be performed. To ensure representative sampling, a sampling plan or protocol is required which ensures that the correct number of samples will be taken at an appropriate frequency and which minimizes sample loss or degradation. A sampling plan should be prepared before the sampling is started to cover the objective, locations and sample size, etc discussed above. The sampling plan should also address the proper use of descriptions, the number, frequency of sampling, decontamination procedures, and handling methods. Depending on whether the waste to be sampled is a liquid, solid, paste, sludge or some combination thereof the sampling methodologies differ. Multiple samples are usually taken in order to determine a statistical average.

Sampling Programme Objective

The objective of sampling is to collect a portion of material small enough in volume to be transported conveniently and handled in the laboratory while still accurately representing the material being sampled. This implies that the relative

Table 13.5: Sampling approach overview.

Sampling Strategy	Definition	Applicability	Advantages/ Disadvantages
Authoritative	Technique where sample locations are selected based on detailed knowledge of the waste stream without regard to randomization.	Waste streams of known physical/chemical properties and concentrations.	Requires in-depth knowledge of properties and constituents of waste streams. Rationale for sample selection must be well documented and defensible.
Random (Simple, Stratified, systematic)	Techniques where sample selection and location are determined through the application of statistical methods.	Used to collect representative samples where data is insufficient to justify authoritative sampling (*e.g.*, waste streams of unknown or variable concentration).	See discussions below for each respective random sampling technique.
Simple Random	All locations/points in a waste or unit from which a sample can be attained are identified, and a suitable number of samples are randomly selected.	Used to collect representative samples of wastes that are heterogeneous throughout the entire waste stream or unit (*e.g.*, multiple drums of unknown origin).	*Advantages*: Most appropriate when little or no information is available concerning the distribution of chemical contaminants. *Disadvantages*: May misrepresent waste streams with areas of high concentration or stratification.
Stratified Random	Areas of non-uniform properties or concentrations are identified and stratified (segregated). Subsequently, simple random samples are collected from each stratum of the waste or unit.	Used to collect representative samples from waste or units that are known to have areas of non-uniform properties (strata) or concentration (hot spots) *e.g.*, surface impoundment with multiple waste layers.	*Advantages*: Provides for increased accuracy of waste streams representation if strata or a typically high or low concentration area is present. *Disadvantages*: Requires greater knowledge of waste stream than for simple random sampling and nay require sophisticated statistical applications.
Systematic Random	The first sampling point is randomly selected but all subsequent samples are collected at fixed space intervals (*e.g.*, along a transect or time intervals).	An alternate procedure used to collect representative samples from modestly heterogeneous waste streams that provide for simplified sample identification.	*Advantages*: Provides for easier sample identification and collection than other techniques. *Disadvantages*: May misrepresent waste streams with unknown areas of high concentration or stratification.

Table 13.6: Major sample types.

Sampling Strategy	Definition	Applicability	Advantages/ Disadvantages
Grab	A sample taken from a Particular location at a distinct point in time.	Most common type used for random sampling. Useful in determining waste streams variability (*e.g.*, range of concentration) when multiple or frequent samples are obtained.	Advantages: Simplest technique, best measure of variability. Disadvantages: May require large number of samples than composting to obtain representative sample.
Composite*	A number of individually collected samples that are combined into a single sample for subsequent analysis.	Used where average or normalized concentration estimates of a waste stream's constituents are desired.	*Advantages*: Reduce analytical costs. May reduce the number of samples needed to gain accurate representation of a waste. *Disadvantages*: Only provides the average concentrations of a waste stream (*i.e.*, information about concentration range is lost).

properties or concentrations of all pertinent component will be the same in the samples as in the material being sampled and that the sample will be handled in such a way that no significant changes in composition occur before the tests are made.

The sample must be representative of the particular substance to be examined and the concentration of the constituent of interest must remain the same until the analytical tests are made. The techniques of sampling vary depending on the type of substances and their uses. Proper location of sampling points and auxiliary equipment is important. In addition, methods of sampling and types of samples are very important.

Number of Samples

Analysis of a large number of samples may, in general, be required to obtain meaningful compositional data since hazardous samples are typically heterogeneous. The number of individual samples that should be analyzed will depend on the kind of information required by the investigation. If an average compositional value is required, a large number of randomly selected samples may be obtained, combined and blended to provide a reasonable homogeneous composite sample from which a sufficient number of sub-samples are analyzed. If composition profiles or the variability of the sample population is of interest, many samples will need to be collected and analyzed individually.

Sampling Equipment

In general, selection of appropriate sampling equipment depends on the physico-chemical properties of the waste. Specific physical parameters affecting this selection

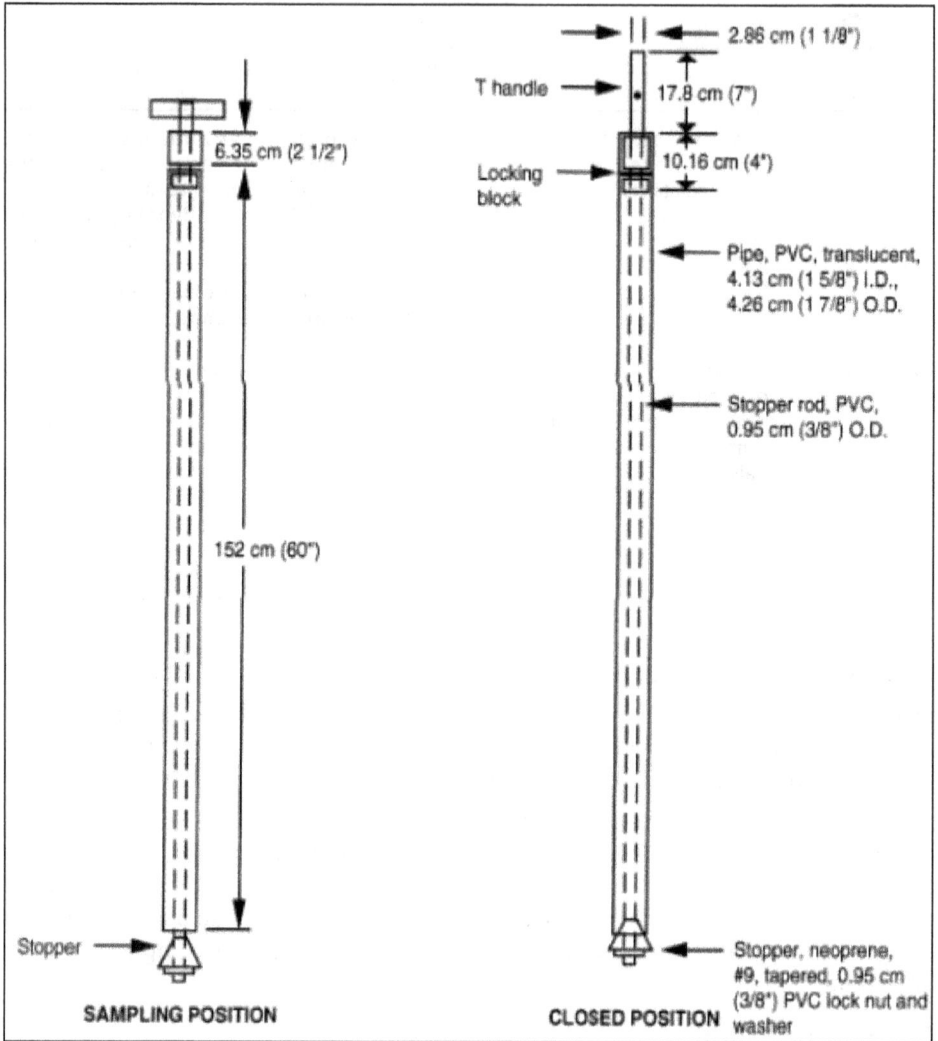

Figure 13.1: Composite liquid waste sampler (Coliwasa).

include whether the wastes are free flowing or highly viscous liquids, crushed, powdered or whole solid matrices, contained in soils, open dumps, etc.

Chemical properties of the waste also significantly influence the selection of equipment. The person collecting the sample should ensure that the sampling equipment is constructed of materials that are not only compatible with wastes, but are not susceptible to reactions that might alter or bias the physical or chemical characteristics of the waste.

Some times, sampling equipment could be waste specific (*e.g.*, oily sludges) or site-specific, *i.e.*, factors such as accessibility to the site, etc. Hence, in such situations, the type of sampling equipment chosen may have to be properly modified for its applications in such situations.

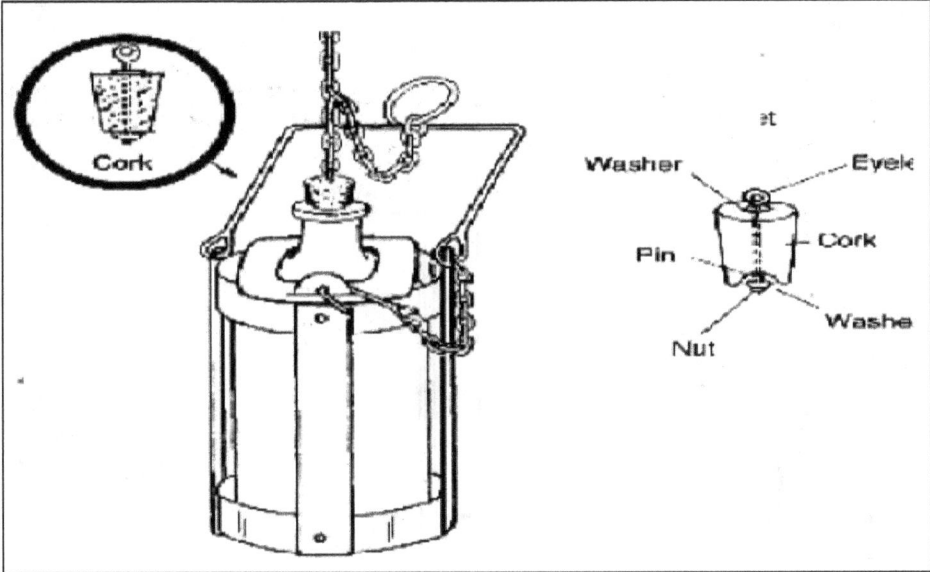

Figure 13.3: Weighted bottle.

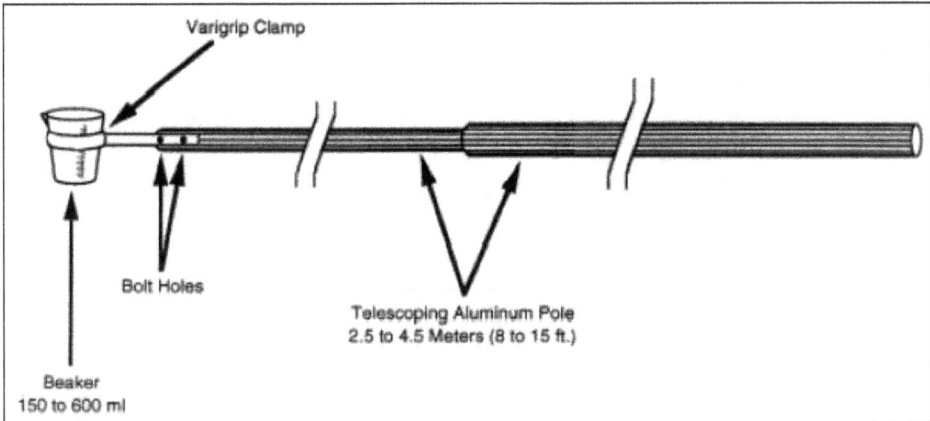

Figure 13.4: Dipper.

Keeping in view of the requirements of sampling strategies along with the physical, chemical, waste and site specific factors associated with the waste to be sampled standard procedures for the sample collection are presented below.

Standard Procedures for Sample Collection (Solid/Semi-Solid/Sludge/Oil)

Samples should be collected under the supervision of a qualified Environmental Engineer, Geologist and concerned Chemist/Scientist.

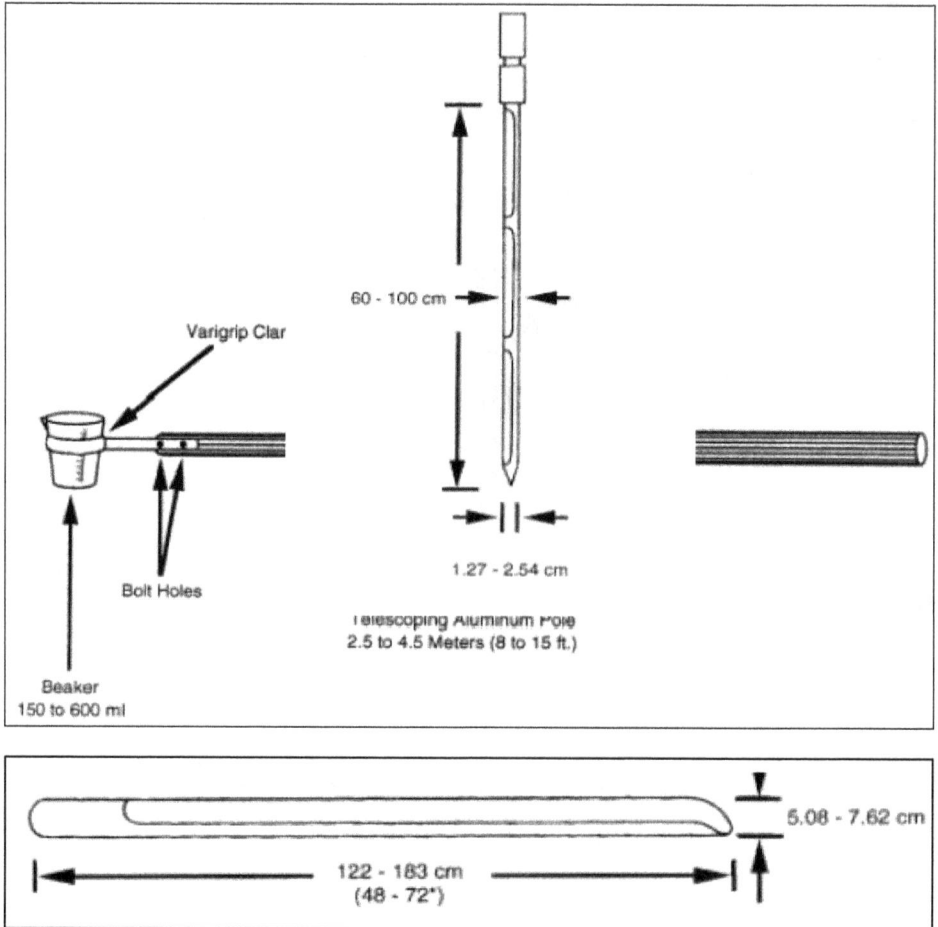

Figure 13.5

Selection of Sampling Approach

Select a suitable sampling approach based on the applicability and associated advantages/disadvantages as detailed in Table 13.5.

Selection of Type of Samples

Decide the types of samples to be collected from different waste streams based on the applicability and associated advantages/disadvantages as detailed in Table 13.6.

* A large number of composite samples or splitting of the composites is recommended for replicate measurement. Cone and quartering is an effective method for homogenizing bulk solid samples. The sample is poured into a cone, which is flattened and quartered. The process is repeated until homogeneity is achieved. The riffle splitter is another commonly used method. The waste is repeatedly poured through the splitter and halves are combined between passes. This method is more efficient with less loss fines and is therefore recommended to homogenize bulk soil samples.

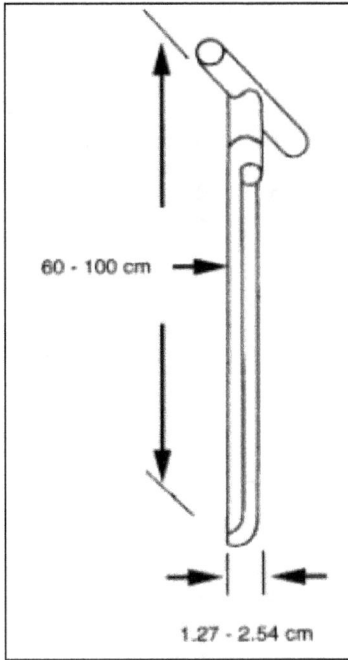

Figure 13.6

60 - 100 cm

1.27 - 2.54 cm

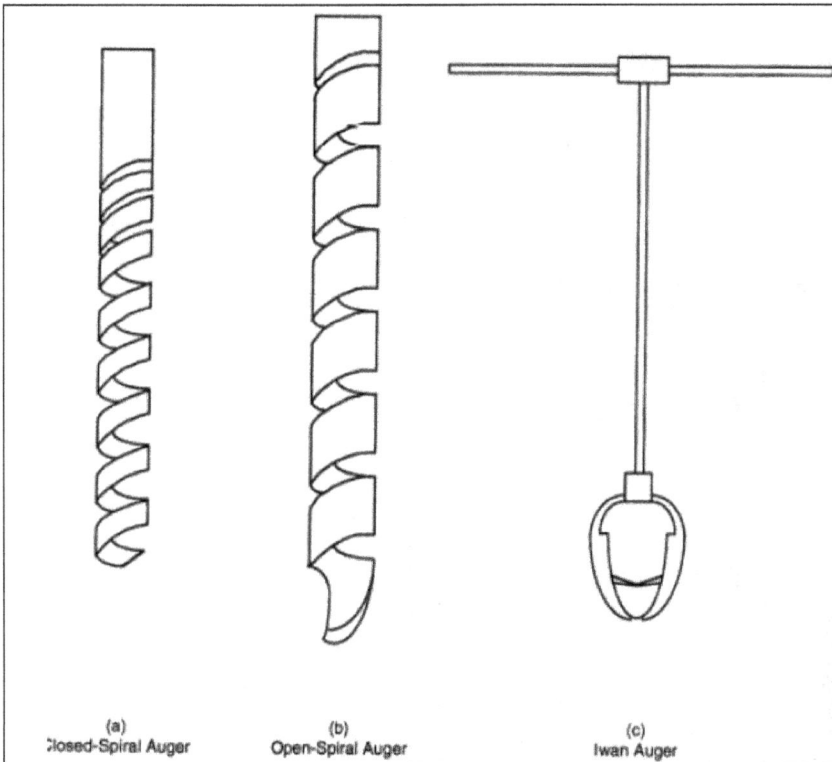

(a)
Closed-Spiral Auger

(b)
Open-Spiral Auger

(c)
Iwan Auger

Figure 13.7

Table 13.7: Applicability of sampling equipment to waste streams, waste location or container

Waste Type	Drum	Sacks and Bags	Open-Bed Truck	Closed Bed Truck	Storage Tanks or Bins	Waste Piles	Ponds, Lagoons and Pits	Conveyor Belt	Pipe
Free-flowing liquids and slurries	Coliwasa	N/A	N/A	Coliwasa	Weighted bottle[a]	N/A	Dipper	N/A	Dipper
Sludges	Trier	N/A	Trier	Trier	Trier	a	A	b	b
Moist powders or granules	Trier	Trier	Trier	Trier	Trier	Trier	Trier	Shovel	Dipper
Dry powders or granules	Thief	Thief	Thief	Thief	a	Trier	Thief	Shovel	Dipper
Sand or packed powders and granules	Auger	Auger	Auger	Auger	Thief	Thief	A	Dipper	Dipper
Large grained solids	Large Trier	Large Trier	Large Trier	Large Trier	Large Trier	Large Trier	Large Trier	Trier	Dipper

a This type of sampling situation can present significant logistical sampling problems, and sampling equipment must be specifically selected or designed based on site and waste conditions. No general statement about appropriate sampling equipment can be made.

Table 13.8: Samplers recommended for various types of waste.

Waste Type	Recommended Sampler	Limitations
Liquids, sludges, and slurries in drums, vacuum trucks, barrels, and similar containers	Coliwasa	Not for containers 1.5 m (5ft) deep
		Not for wastes containing ketones, nitrobenzene, dimethyl formamide or tetrahydrofuran. etc.
		Not for wastes containing hydrofluoric acid and concentrated alkali solutions.
Liquids and sludges in ponds, pits, or lagoons	Plastic Glass Pond	Cannot be used to collect samples beyond 3.5 m (11.5ft). Dip and retrieve sampler slowly to avoid bending the tubular aluminium handle.
Powdered or granular solids in bags, drums, barrels, and similar containers	a) Grain sampler	Limited application for sampling moist and sticky solids with a diameter 0.6 cm (1/4 in.)
	b) Sampling Trier	May incur difficulty in retaining core sample of very dry granular materials during sampling.
Dry wastes in shallow containers and surface soil	Trowel or scoop	Not applicable to sampling deeper than 8cm (3 in.). Difficult to obtain reproducible mass of samples.
Waste piles	Waste pile sampler	Not applicable to sampling solid wastes with dimensions greater than half the diameter of the sampling tube.
Soil deeper than 8cm (3 in.)	a) Soil auger	Does not collect undisturbed core sample.
	b) Veihmeyer sampler	Difficult to use on stony, rocky, or very wet soil.
Wastes in storage tanks	Weighted bottle sampler	May be difficult to use on very viscous liquids.

Selection of the Sampling Equipment

Applicability of various kinds of sampling equipment and recommended sample containers to hazardous wastes for the wastes to be sampled in different environments is presented in Tables 13.7 and 13.8. Figures and dimensions of sampling equipment are presented in Figures 13.2 to 13.7.

A brief description of the sample collection equipment is presented below.

Composite Liquid Waste Sampler (Coliwasa)

The Coliwasa is a device employed to sample free-flowing liquids and slurries contained in drums, shallow tanks, pits, and similar containers. It is especially useful for sampling wastes that consist of several immiscible liquid phases. The equipment consists of a glass, plastic, or metal tube equipped with an end closure that can be opened and closed while the tube is submerged in the material to be sampled (Figure 13.2).

Weighted Bottle

This sampler consists of a glass or plastic bottle, sinker, stopper and a line that is used to lower, raise and open the bottle. The weighted bottle is used for the collection liquids and free-flowing slurries (Figure 13.3).

Dipper

The dipper consists of a glass or plastic beaker clamped to the end of a two- or three-piece telescoping aluminum or fiberglass pole that serves as the handle. A dipper is used for the collection of liquids and free-flowing slurries (Figure 13.4).

Thief

A thief consists of two slotted concentric tubes, usually made of stainless steel or brass. The outer tube has a conical pointed tip that is rotated to open and close the sampler. A thief is used to sample dry granules or powdered wastes whose particle diameter is less than one-third the width of the slots. This equipment can be procured from any suppliers of laboratory items (Figure 13.5).

Trier

A trier consists of a tube cut in half lengthwise with a sharpened tip that allows the sampler to cut into sticky solids and to loosen soil. A trier is used for the collection of moist samples or sticky solids with a particle diameter less than one-half diameter of the trier. Some kinds of triers can be readily procured from the suppliers of the laboratory items and hardware shops or triers can be fabricated as per the required dimensions (Figure 13.6).

Auger

An auger consists of sharpened spiral blades attached to a hard metal central shaft. An auger samples hard or packed solid wastes or soil. Augers can be procured from hardware shops and suppliers of the laboratory items (Figure 13.7).

Scoops and Shovels

Scoops and shovels are used for the collection of granular or powdered material in bins, shallow containers and conveyor belts. Scoops are available at hardware and suppliers of the laboratory items.

Source: USEPA "Samplers And Sampling Procedures For Hazardous Waste Streams, Jan 1980,

Sample Collection Equipment–Other Considerations

The choice of sampling equipment and sample containers will depend upon the previously described waste and site considerations. The following aspects should be carefully examined while suggesting the suitability of sample collection equipment.

The potential interactions between sampling equipment or container material with analytes of interest in the waste to be sampled must be carefully examined and materials that minimizes losses by adsorption, volatilization or contamination must be selected as sampling equipment and sample containers.

Prior to the initial sampling location and between subsequent sampling locations, decontamination procedures should be followed strictly to prevent the introduction of contaminants, *viz.*, negative, positive and cross contamination by the sampling equipment. The decontamination procedure consists of the following steps:

1. Steam clean or scrub all equipment with a non-phosphate detergent.

2. Rinse with tap water.

3. Rinse twice with de-ionized or distilled water.

4. Retrieve the sample, record the details of samples like, quantity, locations, depth, and nearby water bodies to survey the contamination.

5. The minimum sample volume required is specified by the analytical laboratory based on the selected method and required sensitivity of the analysis, and should be verified with the laboratory to ensure that adequate sample volume is obtained.

6. Sample homogenization (*i.e.*, collection of composite samples) should not be performed on samples intended for volatile or semi volatile organics analysis since the mechanical action of mixing exposes a larger surface area of the waste to the air, thus increasing the total amount of volatilization.

Containerization, Preservation and Holding Times of Samples

Selection of appropriate container systems, preservation and transport of samples constitute an important aspect in planning the sampling strategies. The most important factors to consider when choosing containers for hazardous waste samples are compatibility with the waste, cost, and resistance to breakage and volume. Containers must not distort, rupture, or leak as a result of chemical reactions with constituents of waste samples. Thus, it is important to have some idea of the properties and composition of the waste. The containers must have adequate wall thickness to withstand handling during sample collection and transport to the laboratory. Containers with wide mouths are often desirable to facilitate transfer of samples from samplers to containers. Also, the containers must be large enough to contain the optimum sample volume.

Containers for collecting and storing hazardous waste samples are usually made of plastic or glass. Plastics that are commonly used to make the containers include high-density or liner polyethylene, conventional polyethylene, polypropylene, polycarbonate, Teflon FEP (fluorinated ethylene propylene), polyvinyl chloride (PVC) or polymethylpentene. Teflon FEP is almost universally usable due to its chemical inertness and resistance to breakage. However, its high cost severely limits its use. LPE (Liner Polyethylene), on the other hand, usually offers the best combination of chemical resistance and low cost when samples are to be analyzed for inorganic parameters.

Glass containers are relatively inert to most chemicals and can be used to collect and store almost all hazardous waste samples, except those contain strong alkali and hydrofluoric acid. Glass soda bottles are suggested due to their low cost and ready availability. Borosilicate glass containers such as Pyrex and Corex, are more

Table 13.9: Containers, preservation techniques, and holding times for aqueous matrices[A]

Sl.No.	Parameters	Containers	Preservation	Maximum Holding Time
A10	Cyanide, total and amenable to chlorination	P,G	None required	28 days
C12	Nitrates	P, G	Cool to 4°C; If oxidizing agents present add 5 ml of 0.1 n NaAso$_2$ per l or 0.06 g of ascorbic acid per l; adjust pH >12 with 50 per cent NaOH	14 days
C13	Sulfide	P, G	Cool to 4°C, add Zinc acetate	7 Days
A5	Chromium (6)	P, G	Cool to 4°C.	24hours.
A6	Mercury	P,G	HNO$_3$ to pH <2	28 days
A1, A2, A3, A4, A5, A8, A9, B2, B3, B4, B5, B6, B7, B8, B9, B10, C3, C14	Sb, As, Be, Cd, Se, Te, Th, Co, Cu, Pb, Mo, Ni, Sn, W, V, Ag, Ba, Zn	P, G	HNO$_3$ to pH <2	6 Months
A17	Halogenated hydrocarbons	G, PTFE-lined septum	Cool to 4°C.	7 days until extraction.
B14	Nitro and nitroso compounds	G, PTFE-lined cap (Teflone)	Cool to 4°C. 0.008 per cent Na$_2$S$_2$O$_3$[3]	7 days until extraction. 40 days after extraction.
A19	Dieldrin, Aldrin, Endrin	G, PTFE-lined cap	Cool to 4°C	7 days until extraction. 40 days after extraction.
A16	Polychloro Biphenyls	G, PTFE-lined cap	Cool to 4°C	7 days until extraction. 40 days after extraction.
B19	Phenols	G, PTFE-lined cap	Cool to 4°C 0.008 per cent Na$_2$S$_2$O$_3$[3]	7 days until extraction. 40 days after extraction.
A12, A13, A14, A15	PAHs	G, PTFE-lined cap	Cool to 4°c 0.008 per cent Na$_2$S$_2$O$_3$[3]	7 days until extraction. 40 days after extraction.
A17	Halogenated organic compounds	G, PTFE-lined cap	Cool to 4°C 0.008 per cent Na$_2$S$_2$O$_3$[3]	14 days.
	Hydrogen ion (pH)	P, G	None required.	24 hours
	Oil and grease	G	Cool to 4°c	28 days
	Organic carbon, total (TOC)	P, G	Cool to 4°C, store in dark.	28 days

* P: Polyethylene; G: Glass; PTFE: Teflon.

inert and more resistant to breakage than soda glass, but are expensive and not always readily available. Glass or FEP containers must be used for waste samples that will be analyzed for organic compounds.

The containers must have tight, screw-type lids. Plastic bottles are usually provided with screw caps made of the same material as the bottles. Buttress threads are recommended. If the samples are to be submitted for analysis of volatile compounds, the samples must be sealed in airtight containers. Appropriate containers with an appropriate lid type should be used for storing and transporting samples. Refer Table-9 as applicable to the Schedule 2 and 3 of the rules for containers, preservation techniques, and holding times for aqueous matrices.

Samples containing light sensitive organic contaminants should be stored in amber glass bottles with appropriate lids. Samples of an aqueous or solid matrix intended for organic analysis should be stored in glass bottles with appropriate lids or lined caps. In general, a sample size of 1kg of solid is required for a complete comprehensive analysis. If the sample is in the form of sludge or slurry, ensure that the content of solids in the sample collected is at least 1 kg.

Measure/fix the parameters of importance like pH, sulfides, nitrites, at field itself. Preservatives should be added depending upon the parameters at the site itself as indicated at Table 13.9.

Sample Preparation Methods for Analysis of Hazardous Wastes

In this section, sample preparation methods for the analysis of hazardous wastes are presented.

Drying of Samples

Until recently, air-drying was considered to be applicable for most types of analyses. It aids in obtaining a representative analysis portion by producing samples amenable to grinding, sieving and splitting of sample (if required). It is now recognised, however, that even air-drying may modify the chemical form of some species (especially Manganese and Iron) and hence, affect the results obtained. The impact of air drying analysis may be more pronounced in certain sample types. Therefore, air-drying is only applicable to some methods of analysis. It is generally accepted that wastes for metal and some inorganic analytes can be air-dried followed by grinding and sieving.

Sample Preparation

Samples for analysis of metals and other inorganic parameters, some times, requires, removal of extraneous material. A careful judgment by concerned scientist/analyst is often required for the removal of any extraneous material from the sample such as plastics, broken glass pieces, etc. Such situations are normally encountered when the samples are collected from (unauthorized) waste dumpsites in the industrial estates, etc. In general, decisions for removal of any component from the sample must be made on the basis of field information at the time of sample collection and/or the information recorded at the time of sample collection.

The significance of the analyte concentration in the sample or fraction of removed material can then be assessed relative to the entire sample composition. The removed material should be labeled and retained for possible future analysis.

Grinding, Sieving and Homogenization of Sample

Field moist samples will often not be amenable to machine grinding and sieving. For non-volatile analytes in general and most of the inorganic parameters in particular, 50 per cent by weight or 200 g of laboratory sample, whichever is the smaller should be thoroughly ground and mixed in a mortar and pestle to obtain a homogeneous sub-sample. This sub-sample should be stored in a glass screw cap bottle or a suitable container.

All equipment used for this procedure must be cleaned thoroughly with detergent solution, tap water followed by DD Water. The equipment then is dried. Enough care must be exercised to ensure cross-contamination of sample due to equipment used is avoided.

Dry at least 50 per cent by weight or 200 g of sample which ever is smaller by spreading the sample on a shallow tray a suitable non-contaminating material such as plastic or stainless steel. If necessary, break up large clods with a spatula to speed up the drying process. Allow the soils to dry in the air, ideally with the trays placed in a clean air chamber or a non-contaminating oven in the temperature range of 20 - 40 °C. Unless otherwise stated, the analysis portion should pass 2.0 mm or less aperture sieve.

Notes

The process of air drying is applicable only for solid wastes, semi-solid wastes, sludges with no 'free standing liquids' at the time of sample submission at the laboratory.

For slurries with 'free standing liquids' and liquid samples containing low content of solids, the following procedure must be followed.

a) Measure the total volume and weight of slurry or the liquid sample containing solids b) Filter the sample through glass fibre filter paper (0.6 to 0.8 µm) and record the proportions of liquid and solid phases c) Consider filtrate and solid components as separate samples and conduct the analysis. d) Dry the solid component by following the method suggested above (if required) e) Report the concentrations of analytes in the desired format using the data generated as described at Step (a). (*e.g.,* separate results for liquid and solid components or for a combined sample or on dry weight basis)

3. For liquid (non-aqueous) samples, grinding step is not applicable

Caution

Grinding of samples to fine dimensions may produce dust particles, which may present a health hazard. Preparation should be performed in a fume hood and appropriate respiratory protection system conforming to Occupational Health and Safety Standards.

Partitioning of Analysis Portions of Dry Samples

Homogenize the sieved sample by thorough mixing. Homogenization of sieved samples ensures representative analytical data for over all analysis and reproducible results. Divide the homogenized sample using the "Cone-and-Quarter" technique or by any suitable sampling apparatus. The equipment should be made of appropriate material (*i.e.*, stainless steel) to avoid cross-contamination of the sample. Fractions of partitioned sample must be transferred to suitable glass vials/bottles for different inorganic and organic analysis.

Sample Storage

Air-dried or oven-dried samples easily absorb moisture. Immediately after grinding, homogenizing and partitioning, the prepared sample should be transferred into clearly labeled containers having appropriate lid type. Samples should be stored under dry, relatively cool and low light conditions while awaiting analysis.

Note

Hereinafter sample means homogenized and partitioned portion of air-dried or oven dried sample unless otherwise specified.

Estimation of 'Dry Weight' of the Sample

For the purpose of reporting the analytical data, "Dry Weight of the Sample" is defined as "the weight of the sample excluding moisture content only" from the field (moist) sample. This means, weight of all volatile material, including non-aqueous solvents is inclusive of "Dry Weight of the Sample". Hence:

1. When the waste sample is essentially inorganic in nature, weight of the sample obtained after subjecting the sample to 'drying' under the test conditions described in following sections may be taken as 'dry weight of the sample'.

2. In all other cases, moisture content of the sample should be estimated using "Karl-Fisher Titration Method" and correspondingly, 'dry weight of the sample (waste)' must be estimated by deducting the moisture content from the sample.

This requires the estimation of per cent solids and per cent moisture content in the sample estimated at 105°C (oven drying) or per cent Moisture content estimation by Karl Fisher Method (K. F. Method).

Determination of Percent Solids of the Sample

1. Weigh 25-50 g of the field moist sample (on sample as-is-where-is basis) (depending on expected moisture content or per cent solids in the waste) in to a previously weighed or tared crucible. (Select the type of crucible based on the nature and properties of the sample)

2. Dry the sample overnight at 105 °C in a hot air oven
3. Cool crucible in a desiccator for 30 min.
4. Weigh the crucible after cooling
5. Calculate per cent moisture and dry weight of the sample

Determination of Percent Moisture in the Sample

1. Weigh 5-10 g of the field moist sample (on sample as-is-where-is basis) (depending on expected moisture content or per cent solids in the waste) in to a previously weighed or tared crucible. (Select the type of crucible based on the nature and properties of the sample)
2. Dry the sample overnight at 105 °C in a hot air oven
3. Cool the crucible in a desiccator for 30 min.
4. Weigh the crucible after cooling
5. Calculate per cent moisture and dry weight of the sample

Determination of Water in Waste Materials by Karl Fischer Titration

Scope and Application

1. The Karl Fischer titration technique is capable of quantifying the water content of materials from 1 ppm to nearly 100 per cent. Coulometric titration is used for direct analysis of samples with water contents between 1 ppm and 5 per cent, while volumetric titration is more suitable for direct analysis of higher levels (100 ppm to 100 per cent). With proper sample dilution, the range of the coulometric technique can also be extended to 100 per cent water. Both coulometric and volumetric procedures are presented.
2. Multiphasic samples should be separated into physical phases (liquid, solid, etc.) prior to analysis to assure representative aliquots are analyzed.
3. Establishing the water content in a sample may be useful for the reasons to follow.
4. It is useful in determining the total composition of a sample. In combination with other analytical results, the mass balance of a sample can be determined.
5. It is useful in determining the amount of alcohol in an aqueous solution.
6. It is useful when distinguishing an aqueous from a non-aqueous solution.
7. It is useful when setting the proper mixture of feed materials in the incineration of waste.

Principle

For the determination of small amounts of water, Karl Fischer proposed a reagent prepared by the action of sulphur dioxide upon a solution of iodine in a mixture of anhydrous pyridine and anhydrous methanol. Water reacts with this reagent in a two-stage process in which one molecule of iodine disappears for each molecule of

Table 13.10: List of digestion equipment, extraction/distillation apparatus, and instruments required for characterization/analysis of hazardous waste.

Sl.No.	Instruments Required for Analysis and Characterization	Applications
Group-A Digestion Equipment		
1.	Hot plate	Conventional digestion apparatus-All heavy metal digestions can be conducted at 1050°C.
2.	Microwave digestion	This is a sophisticated digestion apparatus. Both closed and open type of digestions can be conducted without any loss of analytes under investigation. Digestion can be done at a time for 12 samples and it is very faster than ordinary digestion.
3.	Digesdhal	Useful for digestion of soil contaminated soil, water wastewater samples with in short periods.
4.	Bomb digestion.	
Group-B Extraction/distillation apparatus		
5.	Soxhlet extraction apparatus.	
6.	Cyanide distillation apparatus	Designed exclusively for reactive cyanide distillation.
7.	Sulfide distillation apparatus.	Designed exclusively for reactive sulfide distillation.
8.	K.D apparatus with Snyder columns	Designed for Pre concentration of the sample with only Solvent evaporation through spiral Snyder column leaving the sample under investigation. Conversion of sample into gaseous phases and back condensation into liquid state takes place.
9.	Fractional distillation apparatus	Distillation of solvents.
10.	Rotavapor.	For solvent recovery.
11.	TCLP Agitator	Shaking or extraction apparatus for leachate generation-Useful for estimation of semi volatile organics and heavy metals from the leachate.
12.	ZHE (Zero Head Space Extractor)	Designed for extraction of leachate without any head space-useful for estimation of volatile organic compounds from leachate.
Group-C Analytical Instruments		
13.	pH meter	Elcrometric measurement of ph.
14.	Karl-fisher moisture content meter	Estimation of moisture content up to 100 per cent in field waste samples.
15.	Conductometer	Measurement of electrical conductivity.
16.	Flash point apparatus	To find flash point of liquid or solvent wastes.
17.	Bomb calorimeter	To find calorific value of the waste.
18.	Uv-Visible spectro-photometer (Preferably Double-beam)	Colorimetric estimation of metals, anoins, phenols etc.

Contd...

Table 13.10–*Contd...*

Sl.No.	Instruments Required for Analysis and Characterization	Applications
19.	Atomic Absorption Spectrophotometer (AAS)	Useful for estimation of trace metal/element analysis. Using this instrument more than 65 elements of the periodic table can be estimated.
20.	AAS/Hydride Vapour Generation system (AAS/GH)	It is an accessory to AAS, and is required to estimate metals like Arsenic, Selenium and Mercury through cold vapor generation.
21.	Elemental Analyser	Useful for estimation of total Organic Carbon, Hydrogen, Nitrogen, Sulphur, and total Oxygen.
22.	Gas Chromatography (GC)	Useful for estimation of semi volatile, and volatile organic constituents present in the waste by specific detectors like FID, FPD, ECD, etc.
23.	Gas Chromatography/ Mass Spectrograph (GC/MS)	It is a coupled technique, of both Gas chromatography with Mass Spectrometer. Initially sample passes through column and gets resoluted, then it gets disintegrated into fragment ions with characteristic M/E ratio. Therefore the resoluted sample can be compared in mass spectrum for further compound confirmation.
24.	Ion Selective Electrodes	It is specific ion meter useful for the quantification of total Sulfide, total cyanide, chloride, nitrate, nitrite and flouride ions. Samples with different type of matrixes can be estimated using this instrument with minimal interferences.
25.	Total Organic Carbon Analyser	Useful for estimation of total organic carbon.

water present. The end point of the reaction is conveniently determined electrometrically using the dead-stop end point procedure. Pyridine free Karl Fisher reagent is now used in view of environmental considerations.

Summary of Method

1. In the volumetric procedure, the sample or an extract of it, is added to a Karl Fischer solvent consisting of sulfur dioxide and an amine dissolved in anhydrous methanol. This solution is titrated with an anhydrous solvent containing iodine. The iodine titrant is first standardized by titrating a known amount of water.

2. In the coulometric procedure, the sample or an extract of it, is injected into an electrolytic cell containing the Karl Fischer solvent, where the iodine required for reaction with water is produced by anodic oxidation of iodide. With this technique, no standardization of reagents is required.

3. In both procedures, the endpoint is determined amperometrically with a platinum electrode that senses a sharp change in cell resistance when the iodine has reacted with all of the water in the sample.

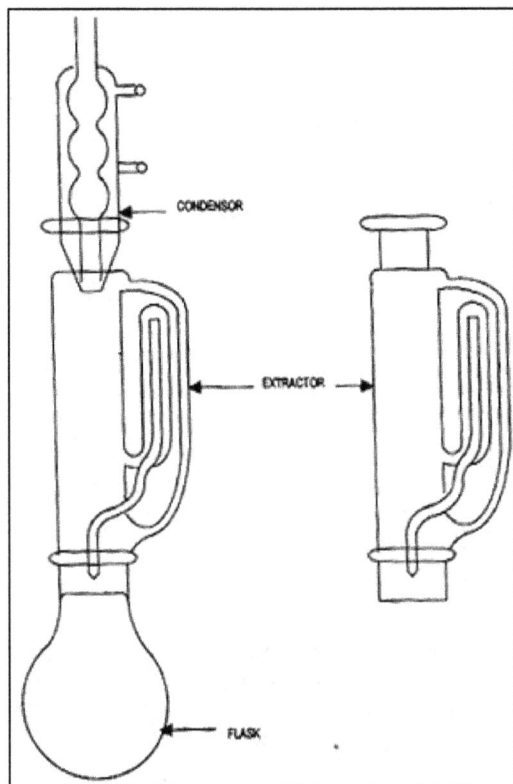

Figure 13.8: Soxhlet extraction apparatus.

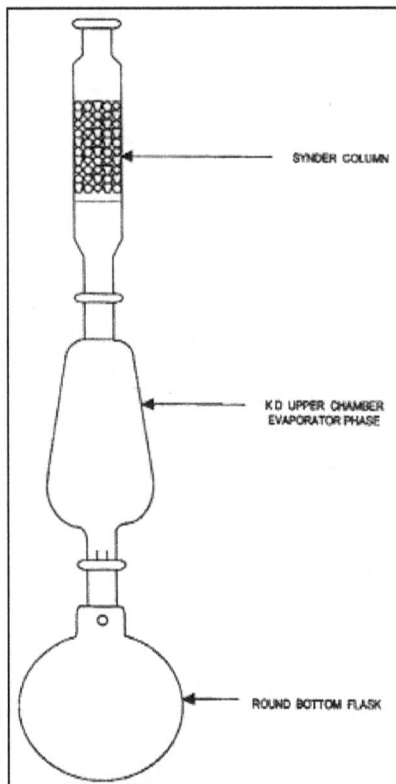

Figure 13.9: Kunderna danish evaporator.

4. In the coulometric procedure, the coulombs of electricity required to generate the necessary amount of iodine are converted to micrograms of water by the instrument microprocessor, while in the volumetric procedure, the volume of iodine titrant required to reach the endpoint is converted to micrograms of water. Most instruments will also calculate concentration (ppm or percent) if the sample weight is keyed in.

Instruments for Analysis and Characterization of Hazardous Waste

The instruments required for carrying out analysis and characterization as per the rules can be classified into 3 groups as follows (Table 13.10).

A. Analytical Instruments

B. Digestion Equipment

C. Extraction/Distillation

A. Analytical Instruments

1. Atomic Absorption Spectrophotometer (AAS): estimation of trace metals in waste samples.

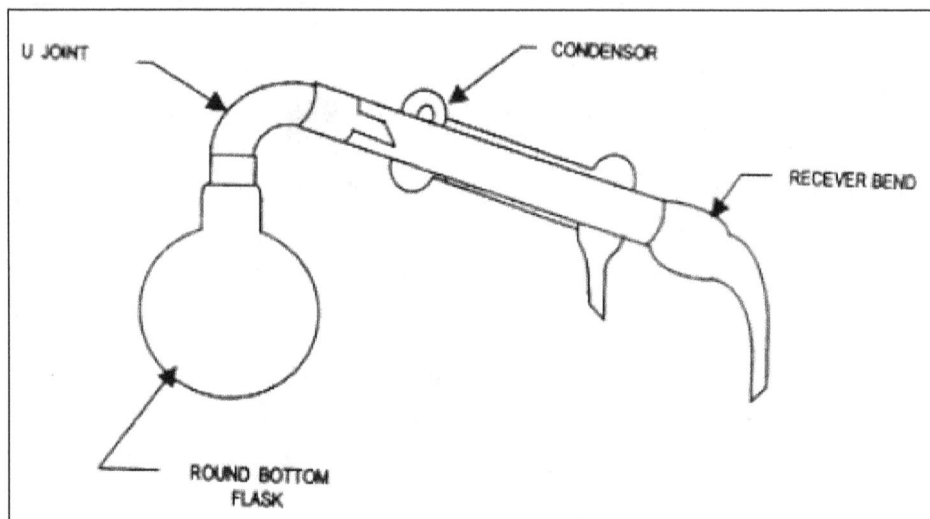

Figure 13.10

2. Gas Chromatography-Mass spectrometer (GC/MS): Estimation of volatile and semi volatile organic compounds. This is a coupled technique used for estimation of organic compounds in which both chromatograms with retention times, and mass/electron ratio will be obtained for exact identification of organic compounds.

3. Gas chromatography: with different detectors like ECD, FID, NPD, and FPD etc.

Apparatus

Soxhlet extractor: 40 mm I.D., with 500 mL round-bottom flask (Figure 13.8).

Kuderna Danish (K-D) apparatus: 10-mL graduated Ground glass stopper to prevent evaporation of extracts (Figure 13.9).

Distillation Apparatus

All glass, consisting of a 1-liter Pyrex distilling apparatus with Graham condenser (Figure 13.10).

Methods for Determination of Hazardous Characteristics of Wastes

Schedule-3 (Part - B) of the rules identified fourteen hazardous characteristics of wastes. Methods for the determination of the listed hazardous characteristics are presented in this section.

Explosives (H1)

An explosive substance or waste is a solid or liquid substance or waste (or mixture of substances or wastes) which is in itself capable by chemical reaction of producing gas at such a temperature and pressure and at such speed as to cause damage to the surroundings.

Flammable Liquids (H3)

The word "flammable" has the same meaning as "inflammable". Flammable liquids are liquids, or mixture of liquids, or liquids containing solids in solution or suspension (for example, paints, varnishes, lacquers, etc., but not including substances or wastes otherwise classified on account of their dangerous characteristics) which give off a flammable vapor at temperatures of not more than 60.5 degrees centigrade, closed-cup test, or not more than 65.6 degrees centigrade, open-cup test.

Flammable Solids (H4.1)

Solids, or waste solids, other than those classed as explosives, which under conditions encountered in transport are readily combustible, or may cause or contribute to fire through friction; self-reactive and related substances which are liable to undergo a strongly exothermic reaction.

Substances or Wastes Liable to Spontaneous Combustion (H4.2)

Substances or wastes which are liable to spontaneous heating under normal conditions encountered in transport, or to heating up on contact with air, and being then liable to catch fire.

Substances or Wastes, in Contact With Water Emit Flammable Gases (H4.3

Substances or wastes, which by interaction with water, are liable to become spontaneously flammable or to give off flammable gases in dangerous quantities.

Oxidizing (H5.1)

Substances or wastes which, while in themselves not necessarily combustible, may, generally by yielding oxygen causes, or contribute to, the combustion of other materials.

Organic Peroxides (H5.2)

Organic substances or wastes that contain the bivalent -O-O- structure are thermally unstable substances which may undergo exothermic selfaccelerating decomposition.

Poisonous (Acute) (H6.1)

Substances or wastes liable either to cause death or serious injury or to harm health if swallowed or inhaled or by skin contact.

Infectious Substances (H6.2)

Substances or wastes containing viable microorganisms or their toxins, which are known or suspected to cause disease in animals or humans.

Corrosives (H8)

Substances or wastes which, by chemical action, will cause severe damage when in contact with living tissues, or, in the case of leakage, will materially damage or even destroy, other goods or the means of transport; they may also cause other hazards.

Liberation of Toxic Gases in Contact With Air or Water (H10)

Substances of wastes, which, by interaction with air or water, are liable to give off toxic gases in dangerous quantities.

Toxic (Delayed or Chronic) (H11)

Substances or wastes which, if they are inhaled or ingested or if they penetrate the skin, may involve delayed or chronic effects, including carcinogen city.

Ecotoxic (H12)

Substances or wastes which if released present or may present immediate or delayed adverse impacts to the environment by means of bioaccumulation and/or toxic effects upon biotic systems.

SETA Flash Closed-Cup Method for Determining Ignitability

Summary of Method

1. By means of a syringe, 2-mL of sample is introduced through a leak proof entry port into the tightly closed Seta flash Tester or directly into the cup, which has been brought to within 30C below the expected flash point.

2. As a flash/no-flash test, the expected flash-point temperature may be a specification (*e.g.*, 600C). For specification testing, the temperature of the apparatus is raised to the precise temperature of the specification flash point by slight adjustment of the temperature dial. After 1 min, a test flame is applied inside the cup and note is taken as to whether the test sample flashes or not. If a repeat test is necessary, a fresh sample should be used.

3. For a finite flash measurement, the temperature is sequentially increased through the anticipated range, the test flame being applied at 50C intervals until a flash is observed. A repeat determination is then made using a fresh sample, starting the test at the temperature of the last interval before the flash point of the material and making tests at increasing 0.50C intervals

Pensky Martens Closed-Cup Method for Determining Ignitability

This uses the Pen sky-Martens closed-cup tester to determine the flash point of liquids including those that tend to form a surface film under test conditions. Liquids containing non-filterable, suspended solids can also be tested using this method.

Summary of Method

The sample is heated at a slow, constant rate with continual stirring. A small flame is directed into the cup at regular intervals with simultaneous interruption of stirring. The flash point is the lowest temperature at which application of the test flame ignites the vapor above the sample.

pH Electrometric Measurement

1. This method is used to measure the pH of aqueous wastes and those multiphasewastes where the aqueous phase constitutes at least 20 per cent of the total volume of the waste.

2. The corrosivity of concentrated acids and bases, or mixed with inert substances, cannot be measured. The pH measurement requires some water content

Summary

The pH of the sample is measured electrometrically using either a glass electrode in combination with a reference potential or a combination electrode. The measuring device is calibrated using a series of standard solutions of known pH.

Apparatus and materials

1. pH meter: Laboratory or field model.
2. Glass electrode.
3. Reference electrode: silver –silver chloride or other reference electrode of constant potential may be used.
4. Magnetic stirrer and Teflon- coated stirring bar.
5. Thermometer and/or temperature sensor for automatic compensation.

Reagents

1. Reagent grade chemicals such as American Chemical Society (ACS) shall be used in all tests. Other grade chemicals can also be used provided the reagent is of sufficiently high purity to permit its use without lessening the accuracy of the determination.

2. Primary standard buffer salts are available from the National Institute of Standards and Technology (NIST) and should be used in situations where extreme accuracy is required. These solutions should be changed at least once in a month.

3. Secondary standard buffers may be obtained from NIST salts or purchased as solvents from commercial vendors. These solutions have been validated by comparison with NIST standards and are recommended for routine use.

Procedure

1. Calibration: Each instrument/electrode system must be calibrated at a minimum of two points that bracket the expected pH of the samples and are approximately three pH units or more apart.

2. For corrosivity characterization, the calibration of the pH meter should include a buffer of pH 2 for acidic wastes and a pH 12 buffer for caustic wastes; also. For corrosivity characterization the sample must be measured at 25 +or-10c if the pH of the waste is above 12.0

3. Place the sample or buffer solution in a clean glass beaker using a sufficient volume to cover the sensing elements of the electrodes and to give adequate clearance for the magnetic stirring bar. If field measurements are being made, the electrodes may be immersed directly into the sample stream to an adequate depth and moved in a manner to ensure sufficient sample movement across the electrodesensing element as indicated by drift-free readings (<0.1 pH).

4. Thoroughly rinse and gently wipe the electrodes prior to measuring pH of samples. Immerse the electrodes into the sample beaker or sample stream and gently stir at a constant rate to provide homogeneity and suspension of solids. Note and record sample pH and temperature.

Repeat measurement on successive aliquots of sample until values differ by <0.1 pH units. Two or three volume changes are usually sufficient.

Toxicity Characteristic Leaching Procedure (TCLP)

1. This method based on USEPA Method 1311-Toxicity Characteristics Leaching Procedure (TCLP), is applicable to the determination of mobility of metals and semi-volatile organic compound in solids.

2. If a total analysis of a solid demonstrates that analyte of interest is not detected, or is present in such low concentrations that regulatory leachate limits cannot be exceeded then it is unnecessary to carry out this leaching test.

Principle

The leaching procedure consists of 3 main steps.

1. Crushing/grinding: The solid sample has to be pass through 9.5-mm sieve. Otherwise, grinding or crushing of the solid is necessary.

2. Determination of appropriate extraction fluid: Depending on the alkalinity of the solid sample, one of two acetic acid leaching fluids is used to extract the soil.

a) Extraction of Solid Sample

1. The solid sample is extracted (20:1 liquid to solid ratio) by shaking it end over end for 18 + or – 2 hours at a controlled temperature. The extract also known as the leachate is then filtered and analysed for the analyte of interest.

2. The moisture content of the solid sample is determined separately and reported with the analytical results.

Summary of the Procedure

From the leachate inorganic and organic species are identified and quantified using appropriate methods as described.

1. The leaching procedure consists of five main steps.

2. For liquid wastes (*i.e.*, those containing 0.5 per cent dry solid material), the waste after filtration through a 0.6 to 0.8 micrometer glass fiber filter, is defined as the TCLP extract.

3. For waste containing greater than or equal to 0.5 per cent solids, the liquid, if any, is separated from the solid phase and stored for later analysis.

4. Particle size reduction: Prior to extraction, the solid material must pass through a 9.5 mm (0.375-in.) standard sieve have a surface area per gram of material equal to or greater than 3.1 cm2, or, be smaller than 1 cm in its narrowest dimension. If the surface area is smaller or the particle size larger than described above, the solid portion of the waste is prepared for extraction by crushing, cutting, or grinding the waste to the surface area or particle size described above. (Special precautions must be taken if the solids are prepared for organic volatile extraction.)

5. Extraction of solid material: The solid material from step 2.2 is extracted for 18 + or − 2 hours with an amount of extraction fluid equal to 20 times the weight of the solid phase. The extraction fluid employed is a function of the alkalinity of the solid phase of the waste. A special extractor vessel is used when testing for volatile organics called Zero Head Space Extractor.

6. Final separation of the extraction from the remaining solid Following extraction, the liquid extract is separated from the solid phase by filtration through a 0.6 to 0.8 micrometer glass fiber filter. If compatible, the initial liquid phase of the waste is added to the liquid extract, and these are analyzed together. If incompatible, the liquids are analyzed separately and the results are mathematically combined to yield a volume-weighted average concentration.

Apparatus

All equipment with which the sample and extract come in contact should be made of inert materials that will not increase or reduce the concentrations of the analytes of interest in the sample. Glass, polytetraflouroethylene, (PTFE) devices may be used when evaluating both organic and inorganic components. Borosilicate glass bottles are recommended in preference to other glass types, especially when inorganics are being evaluated. Vessels made of high-density polyethylene (HDPE), polypropylene (pp) or poly vinyl chloride (PVC) may be used when evaluating only the mobility of metals.

1. Agitation apparatus: The agitation apparatus must be capable of rotating the extraction vessels in an end over-end fashion continuously at 30+ or − 2 rpm.

2. Extraction vessels: Jars or bottles with sufficient capacity to hold the sample and extraction fluid. Two-liter normal capacity bottles are recommended. The vessel type is determined by the analytes of interest.

3. Filtration devices: Filter holders- any filter holder, which meets all of the following requirements. Capable of supporting a 0.6 to 0.8 um glass fibre filter membranes. Has a minimum internal volume of 300 ml (1.5 L

recommended) can hold a filter of minimum size 47 mm in diameter (142 mm filter diameter recommended). Positive pressure filtration units capable of exerting pressures of 350 kpa or more ("commonly called Hazardous Waste Filtration Units"

Note: If the leachate is to be analyzed for metals, the filter must be prewashed with 1 M nitric acid, rinsed with DDW and dried before use. Acid washed filters may also be used for other non-volatile extracts.

4. pH meter: Calibrated to with in + or – 0.05 pH units at 250 °C.

Reagents

All reagents should be of recognized analytical reagent grade.

1. Nitric acid, 1M
2. Hydrochloric acid, 1M
3. Sodium Hydroxide 1 M
4. Glacial acetic acid
5. Extraction fluid

Extraction fluid No.1: Add 5.7 ml of glacial acetic acid to 500 ml DDW, add 64. 3 ml of 1 M NaOH and dilute to 1 liter. The pH of this fluid should be 4.93 + or – 0.05

Extraction fluid No.2: Dilute 5.7 ml of glacial acetic acid to 1 liter. The pH of this fluid should be 2.88 + or – 0.05.

Note: The extraction fluids should be monitored frequently for impurities and the pH checked before use. Discard if impurities are found or pH is not within specifications.

Sample Collection, Preservation and Handling

The quantity of sample needed must be large enough to support all the requirements of this method. There must be sufficient sample to conduct:

1. A preliminary evaluation of appropriate extraction fluid;
2. Determination of particle size reduction; and
3. An extraction of solids for inorganics and or semi-volatile organics.

There should be sufficient sample to provide enough leachate to support each of the necessary analyses, with repeats if necessary and allowing for limits of detection of the analytical methods.

Preservatives must not be added to sample before extraction. Samples should be refrigerated at 40c. The leaching process and analysis of leachate should be carried out as soon as possible. If the extract is to be analyzed for organics, there should be no headspace in the container. Store all extracts at 40c.

Procedure

A minimum of 100 g of sample is required for analysis.

1. *Crushing/grinding*: Examine the sample. The solid has to be able to pass

through a 9.5-mm sieve. Otherwise, grind or crush the solid sample to this size.

2. *Determination of appropriate extraction fluid*:
   (a) Transfer 5.0 g (+ or – 0.1 g) of the sample (<9.5 mm) into a 500 ml beaker or Erlenmeyer flask.
   (b) Add 96.5 ml of DDW to the beaker, cover with a watch glass and stir vigorously for 5 minutes using a magnetic stirrer.
   (c) Measure and record the pH. If the pH is ≤5.0, use extraction fluid No.1.
   (d) If the pH is > 5.0, add 3.5 ml 1 M HCl, slurry briefly, cover with a watch glass, heat to 500C for 10 minutes. Let the solution cool to room temperature and record the pH. If the pH is ≤ 5, use extraction fluid No.1. Otherwise, use extraction fluid No.2.

3. *Extraction of solid waste*: Enough solid should be used for the extraction (20:1 liquid to solid ratio) such that the volume of leachate will be sufficient to support all of the analyses required. If the volume of leachate from a single extraction is insufficient, several extractions may be performed and the extracts combined for analysis.

   A reagent blank with no solid sample should be included with each process batch of samples.

   (a) Weigh at least 100 g (+ or-0.1 g) of the field sample (<9.5 mm) into an extraction vessel.
   (b) Add an amount of the appropriate extraction fluid equivalent to 20 times the weight of the sample, to the extraction vessel.
   (c) Close the extraction vessel tightly (it is recommended that Teflon tape be used to ensure a tight seal), secure in the rotary agitation device and rotate at 30 + or – 2 rpm for 18+ or – 2 hours at ambient temperature (23+ or – 20 °C).

   Note: For some types of solid sample, during the agitation, pressure may build up with in the extraction vessel eg. From evolution of gasses. At periodic intervals (*e.g.* After 15 minutes, 30 minutes and 1 hour) this pressure should be released in fume hood.

   (d) Assemble the filter holder and filter following manufacturers instructions.
   (e) Place the 0.6 to 0.8 μm glass fibre filter on the support screen and secure.
   (f) At the end of the extraction period, transfer the sample to the filter holder and filter the sample.
   (g) Seal the filtration device and gradually apply vaccum or gentle pressure of 7-70 kPa, and if no additional liquid has passed through the filter in any 2-minute interval, slowly increase the pressure in 70kPa increments to a maximum of 350 kPa. Repeat this until pressurizing gas begins to move through the filter or when liquid flow has ceased at 350kPa. *I.e.*filtration does not result in any additional filtrate within any 2-minute period.

(h) The glass fibre filter may be changed, if necessary, to facilitate filtration. The filtrate collected is called the leachate.

4. *Analysis of leachate*

(a) Analyze the leachate as soon as possible. If the analysis cannot be carried out immediately, transfer suitable volumes of the extract into appropriate containers. If the extract is to be analyzed for organics, there should be no headspace in the container. Store all extracts at 40°C.

(b) Report the analyte concentrations of the leachate in mg/L.

(c) Determine the moisture content of the soil separately and report it together with the analytical results.

Quality Control

1. A minimum of one blank must be performed for process batch, using the extraction fluid and extraction vessel type used for the samples.

2. A matrix spike should be performed for each solid sample type within each batch. Matrix spikes are to be added to the leachate and before preservation. Matrix spikes should not be added prior to extraction of the sample.

Modified Test for Shake Extraction of Solid Waste with Water

Materials

1. Dilution water double distilled water adjusted to pH 5.5 by acetic acid buffer acid addition immediately to use.

2. Tailings solids should be representative of the project material.

Apparatus

1. Agitator may of any type that meets the general requirement at ASTM D3957.

2. Containers for agitating slurry may be at any size compatible with the agitator apparatus and shall be clean, new, acid washed and rinsed before use.

Procedure

1. Prior to selecting tailing solids for the procedure they shall be well mixed and tree of lumps or standing water. Keep the materials in pails, firmly closed when not in use and store in a cool location.

2. Immediately after selection take a weighed quantity at well-mixed tailings material and calculate the dry weight of solids and the weight of associated waterfront the given solids content.

3. Add sufficient freshly prepared dilution water to give a 4:1 liquid to solid ratio (solids content of 20 per cent by weight). *i.e.* (dry weight of solids in slurry)/(water in slurry + water added) x 100 = 20 percent.

4. Sufficient material should be prepared to almost fill the containers to be placed on the agitator.

5. Place on agitator apparatus and run for 18 hours.

6. After the completion of the agitation remove the surface water by decantation followed by pressure filtration or centrifugation (good water recovery is essential).

7. Store all the water recovered without preservation or filtration in a single large clean carboy or else add equal proportions of each new batch or aqueous solution generated to all leach ate storage containers used.

8. Where preservation is required for analysis representative sub-samples should be taken and suitably treated and labeled.

Hazardous Waste Treatment

The various options for hazardous waste treatment can be categorised under physical, chemical, thermal and biological treatments.

Physical and Chemical Treatment

Physical and chemical treatments are an essential part of most hazardous waste treatment operations, and the treatments include the following:

1. Filtration and Separation

Filtration is a method for separating solid particles from a liquid using a porous medium. The driving force in filtration is a pressure gradient, caused by gravity, centrifugal force, vacuum, or pressure greater than atmospheric pressure. The application of filtration for treatment of hazardous waste fall into the following categories:

☆ *Clarification*, in which suspended solid particles less than 100 ppm (parts per million) concentration are removed from an aqueous stream. This is usually accomplished by depth filtration and cross-flow filtration and the primary aim is to produce a clear aqueous effluent, which can either be discharged directly, or further processed. The suspended solids are concentrated in a reject stream.

☆ *Dewatering* of slurries of typically 1 per cent to 30 per cent solids by weight. Here, the aim is to concentrate the solids into a phase or solid form for disposal or further treatment. This is usually accomplished by cake filtration. The filtration treatment, for example, can be used for neutralisation of strong acid with lime or limestone, or precipitation of dissolved heavy metals as carbonates or sulphides followed by settling and thickening of the resulting precipitated solids as slurry. The slurry can be dewatered by cake filtration and the effluent from the settling step can be filtered by depth filtration prior to discharge.

2. Chemical Precipitation

This is a process by which the soluble substance is converted to an insoluble form either by a chemical reaction or by change in the composition of the solvent to diminish the solubility of the substance in it. Settling and/or filtration can then remove the precipitated solids. In the treatment of hazardous waste, the process has a wide applicability in the removal of toxic metal from aqueous wastes by converting them to an insoluble form. This includes wastes containing arsenic, barium, cadmium, chromium, copper, lead, mercury, nickel, selenium, silver, thallium and zinc. The sources of wastes containing metals are metal plating and polishing, inorganic pigment, mining and the electronic industries. Hazardous wastes containing metals are also generated from cleanup of uncontrolled hazardous waste sites, *e.g.*, leachate or contaminated ground water.

3. Chemical Oxidation and Reduction (Redox)

In these reactions, the oxidation state of one reactant is raised, while that of the other reactant is lowered. When electrons are removed from an ion, atom, or molecule, the substance is oxidised and when electrons are added to a substance, it is reduced. Such reactions are used in treatment of metal-bearing wastes, sulphides, cyanides and chromium and in the treatment of many organic wastes such as phenols, pesticides and sulphur containing compounds. Since these treatment processes involve chemical reactions, both reactants are generally in solution. However, in some cases, a solution reacts with a slightly soluble solid or gas.

There are many chemicals, which are oxidising agents; but relatively few of them are used for waste treatment. Some of the commonly used oxidising agents are sodium hypochlorite, hydrogen peroxide, calcium hypochlorite, potassium permanganate and ozone. Reducing agents are used to treat wastes containing hexavalent chromium, mercury, organometallic compounds and chelated metals. Some of the compounds used as reducing agents are sulphur dioxide, sodium borohydride, etc. In general, chemical treatment costs are highly influenced by the chemical cost. This oxidation and reduction treatment tends to be more suitable for low concentration (*i.e.*, less than 1 per cent) in wastes.

4. Solidification and Stabilisation

In hazardous waste management, solidification and stabilisation (S/S) is a term normally used to designate a technology employing activities to reduce the mobility of pollutants, thereby making the waste acceptable under current land disposal requirements. Solidification and stabilisation are treatment processes designed to improve waste handling and physical characteristics, decrease surface area across which pollutants can transfer or leach, limit the solubility or detoxify the hazardous constituent. To understand this technology, it is important for us to understand the following terms:

 ☆ *Solidification*: This refers to a process in which materials are added to the waste to produce a solid. It may or may not involve a chemical bonding between the toxic contaminant and the additive.

 ☆ *Stabilisation*: This refers to a process by which a waste is converted to a more chemically stable form. Subsuming solidification, stabilisation represents the use of a chemical reaction to transform the toxic component to a new, non-toxic compound or substance.

★ *Chemical fixation*: This implies the transformation of toxic contaminants to a new non-toxic compound. The term has been misused to describe processes, which do not involve chemical bonding of the contaminant to the binder.

★ *Encapsulation*: This is a process involving the complete coating or enclosure of a toxic particle or waste agglomerate with a new substance (*e.g.*, S/S additive or binder). The encapsulation of the individual particles is known as micro-encapsulation, while that of an agglomeration of waste particles or micro-encapsulated materials is known as macro-encapsulation.

In S/S method, some wastes can be mixed with filling and binding agents to obtain a dischargeable product. This rather simple treatment can only be used for waste with chemical properties suitable for landfilling. With regard to wastes with physical properties, it changes only the physical properties, but is unsuitable for landfilling. The most important application of this technology, however, is the solidification of metal-containing waste. S/S technology could potentially be an important alternative technology with a major use being to treat wastes in order to make them acceptable for land disposal. Lower permeability, lower contaminant leaching rate and such similar characteristics may make hazardous wastes acceptable for land disposal after stabilisation.

References

1. www.envfor.nic.in/legis/hsm/HAZMAT_2265_eng.pdf "Hazardous Waste (Management, Handling and Transboundary Movement) Rules, 2008" and subsequent amendments 2009.

2. Protocol for Performance Evaluation and Monitoring of the Common Hazardous Waste Treatment Storage and Disposal Facilities including Common Hazardous Waste Incinerators, HAZWAMS/......../2009-2010

3. CPCB, October, 2009.

4. "Guidelines for Storage of Incinerable Hazardous Wastes by the Operators of Common Hazardous Waste Treatment, Storage and Disposal Facilities and Captive HW Incinerators, HAZWAMS/.../2005-2006, CPCB, November, 2008.

5. Methods and Standard operating procedures (SOPs) of Emission Testing in Hazardous Waste Incinerator, LATS, CPCB, September 2007.

6. National Inventory of Hazardous Wastes Generating Industries and Hazardous Waste Management in India, CPCB, February, 2009.

Strategic Approaches

14

Community Based Risk and Vulnerability Assessment: Approach and Challenges

Pallavee Tyagi and Anil Kumar Gupta

Introduction

Over the past few decades, there was an exponential increase in human and material losses from disaster events, though there was no clear evidence that the frequency of extreme hazard events had increased. This indicated that the rise in disasters and their consequences was related to a rise in people's vulnerability, induced by human-determined paths of development. An evolution in approaches from relief and response to vulnerability analysis to risk management – has started influencing how disaster management programs are now being planned and financed. From this realization that people's vulnerability is a key factor determining the impact of disasters on them, emphasis shifted to using "vulnerability analysis" as a tool in disaster management.

As mentioned earlier, a broad consensus is emerging in favour of community-based Disaster Risk Reduction (DRR) approaches, since it is at community level that physical, social and economic risks can be adequately assessed and managed. Over the last decade, growing recognition of the necessity of enhanced community participation for sustainable disaster reduction has often been translated into some actions to carry out community-based vulnerability and risk assessments.

A good example of NGO involvement in this area is the development of specific toolkits for participatory vulnerability and risk assessment and similar activities.

These tools are aimed at assisting field workers and communities to analyze people's vulnerability, draw up action plans, mobilise resources and enact appropriate policies, laws and strategies to reduce community vulnerability to disasters.

Such approaches are based on the idea that communities know their own situations best and that any analysis should be built on their knowledge of local conditions. The approaches seek to use the outputs of local-level analysis to inform national and international level action and policies. Ideally, they should also empower communities to take charge of their own efforts to identify and address vulnerability, and enable them to find opportunities to enhance their resilience to natural hazards.

Quantification of Risk

Vulnerability and risk analysis provides a structured analytical procedure to identify and quantify hazards and to estimate the probability and consequences of their occurrence. It must be emphasized that the absolute risk is a complex, multiplicative function of the hazard level and the vulnerability of a community. In an illustrative sense, this means that:

$$\text{Disaster Risk} = \frac{(\text{Hazard} \times \text{Vulnerability})}{\text{Capacity}}$$

Risk consists of hazard and vulnerability. We can define "Hazard" like "A threatening event, or the probability of occurrence of a potentially damaging phenomenon within a given time period and area" (European Centre of Technological Safety, 2000). When a hazardous event occurs, the damage depends on the elements at risk like population, buildings and civil works, economic activities, public services and infrastructure, etc. exposed to hazards" (European Centre of Technological Safety, 2000).

Vulnerability is a complex notion defining the resulting impacts for each hazard. In general "Vulnerability" is the "Susceptibility to damage from adverse factors or influences" (Regional Vulnerability Assessment of United States Environmental Protection Agency.

Disaster Risk Assessment

Assessment is a process (usually undertaken in phases) of collecting, interpreting and analyzing information from various sources. Risk assessment is an integral component of the process by which individuals, communities and societies cope with hazards. Traditionally it has been done by economists, scientists and insurers, government agencies on agriculture, environmental management, health, public works and highways, etc. who are concerned with estimating probable damages and proposing mitigation measures based on cost-benefit analysis. The outputs of community-based approaches to disaster risk assessments are quantitative estimates of probable loss of life, damage to property and the environment, based on the criteria developed by the community. The risk measurement is then summarized as severe, moderate and minor or high, medium and low.

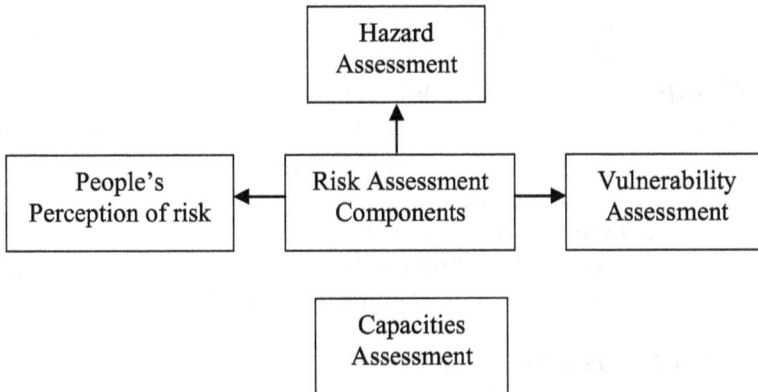

Figure 14.1: Risk assessment components.

Community-Based Disaster Risk Assessment

As community is the key actor as well as the primary beneficiary of disaster risk reduction, it is important to know who in the community should be involved. The most vulnerable are the primary actors in a community. The focus should be at the household level. As all individuals, houses, organizations and services stand a chance of being affected they should all be involved for effective

CBDM. But before working on disaster risk reduction, differing perceptions, interests, and methodologies have to be recognized and a broad consensus on targets, strategies and methodologies have to be reached. To enrich the community's involvement in risk reduction it is important to first assess the risk with the help of the community.

Community Risk Assessment is a participatory process of determining the nature, scope and magnitude of negative effects of hazards to the community and its households within an anticipated time period. It determines the probable or likely negative effect (damage and loss) on 'elements at risk' (people - lives and health; household and community structures, facilities and services – (houses, schools, hospitals, etc.); livelihood and economic activities (jobs, equipment, crops, livestock, etc.); lifelines – (access roads and bridges). Why particular households and groups are vulnerable to specific hazards and why others are not are also analyzed. The coping mechanisms and the resources (capacities) present in the community are also essential considerations in community risk assessment. Participation of community members is an essential component of community based risk assessment, which determines the methodologies and tools used. Community risk assessment combines both scientific and empirical data concerning known hazards and other possible threats to the community. There are specific tools and methods that can make the process of community risk assessment most effective, as following:

- ☆ *Review of secondary information:* collection of relevant information from published or unpublished sources and past lessons of community.
- ☆ *Direct observation:* systematic observation of people and relationships, objects, events, processes and recording these observations to get a better picture of the community

☆ *Semi-structured interview:* informal discussions with the community members using a flexible guide of questions – interviews, group discussions or bunch of people sitting around the table (BOPSAT)

☆ *Drama, Role-play and Simulations* by acting out a particular situation

☆ *Diagramming and visualization tools* by drawing maps, diagrams, etc. to illustrate, analyze, make relations or draw trends. Historical profile, mapping, modeling, transect, seasonal calendar, institutional and social network analysis, livelihood/class analysis, problem tree, gendered resource mapping are some examples of diagrammatic tools

Components of Risk Assessment

Community based risk assessment has four main inter-related steps (Figure 14.1). These are:

1. Hazard Assessment

It determines the likelihood of experiencing any natural or human-made hazard or threat in the community. Assessment includes the nature and behavior of each of the hazards the community is exposed to. Hazards may be of following types:

(a) *Naturally caused* is result from the forces of nature. These may be due to severe weather conditions *e.g.* extreme temperatures, fog, hailstorm, lightning storm, hurricane, windstorm, tornado, ice/sleet storm, snowstorm/blizzard; Geological reasons *e.g.* earthquake, landslide, land

---

One should ask the following questions to explore the physical, social and motivational vulnerability of an area

Physical/material vulnerability

What productive resources, skills and hazards exist?

What made the people affected by disaster physically vulnerable: was it their economic activities (*e.g.* farmers cannot plant because of floods), geographic location (*e.g.* homes built in cyclone-prone areas) or poverty/lack of resources?

Social/organizational vulnerability

What are the relations and organization among people?

How society is organized, its internal conflicts and how it manages them are just as important as the physical/material dimension of vulnerability?

What the social structure was before the disaster and how well it served the people when disaster struck; one can also ask what impact disasters have on social organization?

Motivational/attitudinal vulnerability

How does the community view its ability to create change?

What people's beliefs and motivations are, and how disasters affect them?

Table 14.1: Vulnerability assessment factors and indicators.

Risk Component	References	Factors	Indicators
Vulnerability	Risk Management	Resilience factors	Comprehensive Risk Assessment and support/system
(Resilience)			City's Disaster Management Plan defining roles and responsibility relationship and linking
			Land-use planning
			Building regulations
			Regulation control and enforcement
			Emergency Operating Centers
			Forecast warning and monitoring system
			Community level preparedness in high risk management
			Public awareness of Risk
			Education and training
			Involvement of private sector, NGOs
			Insurance protection
			Business continuity
Vulnerability	Direct impacts	Physical	Population (density distribution)
(Exposure)			Natural resources
			Housing distribution
			Public buildings
			Critical facilities:
			- Basic urban services (waters, waste...)
			- Emergency facilities
			- Buildings of high occupancy
			- Schools and other educational buildings
			- Key economic sectors
			- Communication facilities
			- Transportation facilities
			- Cultural and historic patrimony.
	Indirect effects	Socio-economic	High Risk groups
	Intangible **losses**		Family income
			Loss of production and employment
			Loss of collective services

subsidence or Hydrological based eg. erosion, floods, drought Others: wild fires, agriculture and food human, water quality, human health emergencies and epidemics.

| Risk Assessment | Hazard Assessment | • Define hazards and its types.
• Identifies areas in the country where the hazard is likely to be a problem.
• Analyse and determined which hazards are most likely to affect a given community.
• Data analysis attempts to predict the nature, frequency, and intensity of future hazards; the area most likely to be affected; and the onset time and duration of future events.

Note: Discuss past events as examples. |
| | Vulnerability Assessment | • Define Vulnerability and its components.
• Analyse of the information helps determine who is most likely to be affected, what is most likely to be destroyed or damaged.
• Mapping and assessment of social, economic and environmental vulnerabilities of populations.

Note: Demonstration of any case study to analyse physical, social, economic and environmental vulnerability to hazards at the local level. |

| Capacity assessment | • Define the people's coping capacities (critical facilities) with specific and multiple disasters.
• Analyse the resources available and their status for preparedness, mitigation and emergency responses; who has access to and control over these resources.
• Identify the perception of risks of the heterogeneous groups and sectors, which make up the community; measurement of the community's disaster risks based on people's perception.
Note: Evaluate the coping capacity at given site (information can be obtained from government officials, public utility companies, health departments, hospital associations, and school authorities) |
| Mitigation Options | • Discuss various mitigation options for particular hazard.
• Identify and listing of most suitable and practical options.
• Discuss resources that are locally available for specific mitigation.
• Identification of external resources required. |

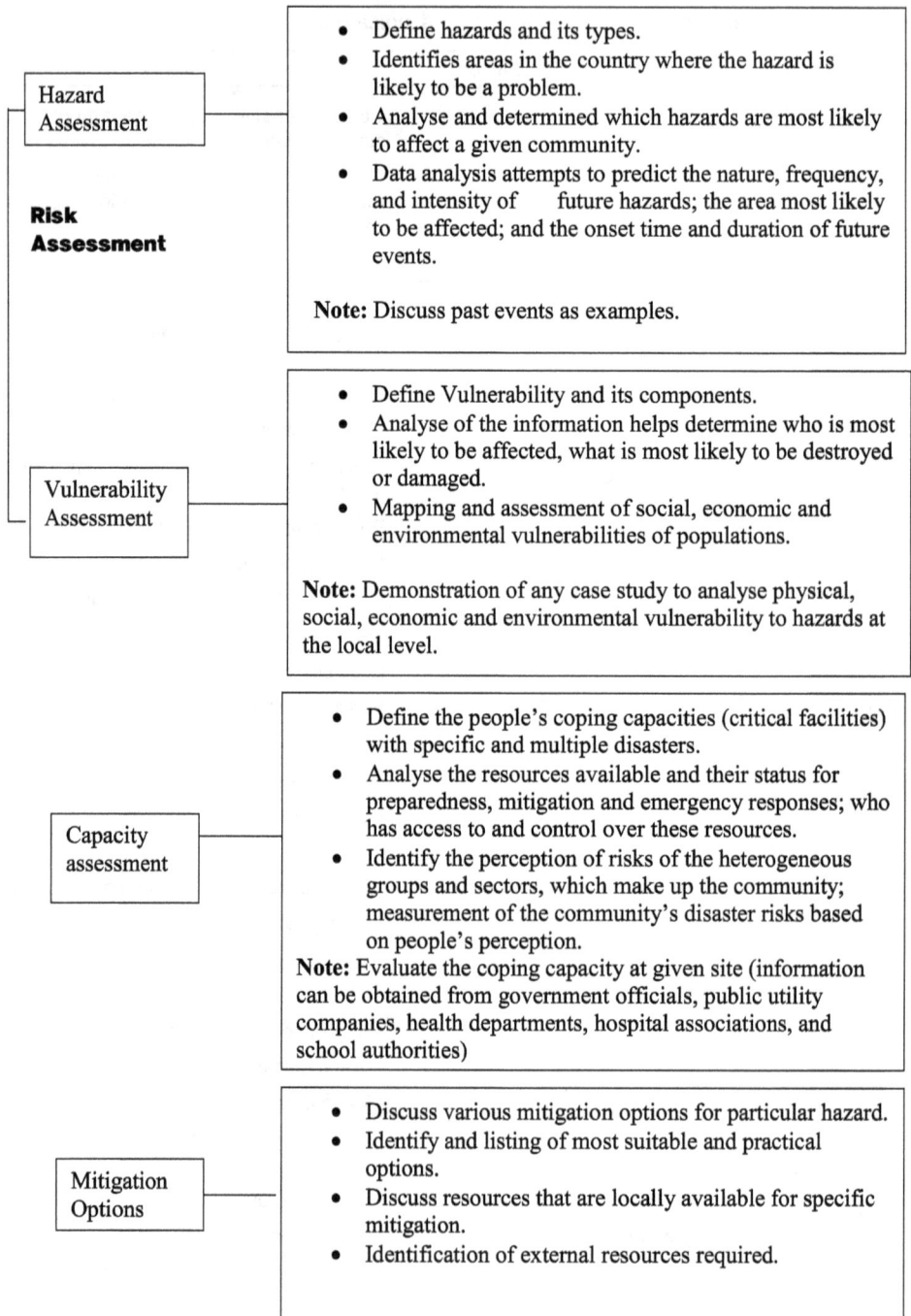

Figure 14.1: Tools Points to be described by Trainer.

(b) *Technological* are a wide range of conditions emanating from the manufacture, transportation, and use of modern technology and substances such as chemicals, explosives, flammables and radioactive materials *e.g.* explosions/fires, building/structural collapse, critical infrastructure failure.

(c) *Human-caused* are due to direct impact of human activities *e.g.* Special events, Civil disorders, Sabotage, Terrorism.

2. Vulnerability Assessment

It identifies what elements are at risk and why they are at risk that means unsafe conditions resulting from dynamic pressures which are consequences of root or underlying causes. Vulnerability is a term used to describe exposure to hazards and shocks. Many of our community's most vulnerable populations are often those most at risk, and those most deeply affected by disaster. While vulnerability can be described in many ways, consider the following three descriptions (a) inability to gain *access, synthesize and use* hazard reduction information to protect lives, homes and businesses, (b) inability to take *steps* to reduce vulnerability to hazards before an event, or take protective steps during or after an event, (c) incapable of *absorbing* the effects of a disaster in the long term. Vulnerability defines the characteristics of a person or group and their situation that influence their capacity to anticipate, cope with, resist and recover from the impact of a hazard (Wisner *et al.*, 2004). It involves a combination of factors that determine the degree to which someone's life, livelihood, property and other assets are put at risk by a discrete and identifiable event (or series or cascade of such events) in nature and society. To analyse and viewing the people's vulnerability, it is categorized under three broad, interrelated areas: physical/material, social/organisational and motivational/attitudinal.

Each of the three categories comprises a wide range of features:

(a) *Physical/material vulnerability:* The most visible area of vulnerability is physical/material poverty. It includes land, climate, environment, health, skills and labour, infrastructure, housing, finance and technologies. Poor people suffer from crises more often than people who are richer because they have little or no savings, few income or production options, and limited resources. They are more vulnerable and recover more slowly.

(b) *Social/organizational vulnerability:* This aspect includes formal political structures and the informal systems through which people get things done. Poor societies that are well organized and cohesive can withstand or recover from disasters better than those where there is little or no organization and communities are divided (*e.g.* by race, religion, class or caste).

(c) *Motivational/attitudinal vulnerability:* This area includes how people in society view themselves and their ability to affect their environment. Groups that share strong ideologies or belief systems, or have experience of co-operating successfully, may be better able to help each other at times of disaster than groups without such shared beliefs or those who feel fatalistic or dependent. Crises can stimulate communities to make extraordinary efforts.

Risk refers to Natural and induced Hazards, Exposure and Vulnerability. It is measured through a direct Impact assessment (physical damages), Indirect Effects (socio-economic and functional consequences) and Intangible Losses assessment, complemented by a Resilience Factors analysis based on Risk Management Practices shown in Table 14.1.

3. Capacities Assessment

identifies the people's coping strategies; resources available for preparedness, mitigation and emergency response; who has access to and control over these resources.

4. People's Perception of Risk

Identifies the perception of risks of the heterogeneous groups and sectors, which make up the community; measurement of the community's disaster risks based on people's perception.

Risk and Vulnerability Assessment Methodology

This process has four sequential stages, which can be operationalised before a disaster occurs to reduce future risks. Each stage grows out of the preceding stage and leads to further action. Together, the sequence can build up a planning and implementation system, which can become a powerful disaster risk reduction tool. The stages in risk and vulnerability assessment are as follows:

References

A System Vulnerability Visualization Architecture, U.S. Government Program Sponsor: Air Force Research Laboratory/IFGB.

Assessment of vulnerability to flood impacts and damages, 2001, Disaster Management Programme, UNCHS (Habitat).

B. Barroca, P. Bernardara, J. M. Mouchel, and G. Hubert, 2006, Indicators for identification of urban flooding vulnerability, Natural Hazards and Earth System Sciences, 6, 553-561.

Dr. Yodamani Suvit, Disaster Risk Management and Vulnerability Reduction: protecting the Poor, Paper Presented at The Asia and Pacific Forum on Poverty organized by the Asian Development Bank.

Fabien Nathan, 2005, Vulnerabilities to Natural Hazard: Case Study on Landslide Risks in La Paz, Paper for the World International Studies Conference (WISC) at Bilgi University, Istambul, Turkey.

NGOs and Disaster Risk Reduction: A Preliminary Review of Initiatives and Progress Made, 2006, Background Paper for a Consultative Meeting on a "Global Network of NGOs for Community Resilience to Disasters" Geneva.

Ronda R. Henning and Kevin L. Fox, 1999, The Network Vulnerability Tool (NVT).

Terry cannon, John Twigg and Jennifer Rowell, 1997, Social Vulnerability, Sustainable Livelihoods and Disasters, Report to DFID, Conflict and Humanitarian Assistance Department (CHAD) and Sustainable Livelihoods Support Office.

training.fema.gov/EMICourses/E464CM/02 per cent 20Unit per cent 202.pdf

www.adpc.net

www.cis.stclaircounty.org/downloads/chap4_pp131137_draft.pdf

www.dhs.vic.gov.au

www.medscape.com/viewarticle/513260

www.noaa.gov

www.bom.gov.au/bmrc/basic/org_hp.htm

www.iisd.org/pdf/2007/climate_bg_vulnerability_adap.pdf

www.unisdr.org/eng/about_isdr/basic_docs/LwR2004/ch2_Section3.pdf

15

Toxicological Approach to Environmental Risk Management

Pallavi Saxena and Anil K. Gupta

Introduction

Toxicology is the fundamental science of poisons dealing with the study of the adverse effects of chemicals on living organisms (Ali, 1993). Hence, toxicology plays an important role with respect to the environment in many ways. The risks that chemicals pose to man kind and environment have become the subject of major scientific, social, economic and political concern in recent decades (Pandya, 1993) Many of the industrial chemicals, petroleum hydrocarbons affect the health of human beings, animals and cause ecological problems, besides the occupational disorders among the workers (Browning, 1987). The health risks associated with the increasing use of chemicals have to be assessed and abated by suitable anticipatory action. There is also wide range of xenobiotics to which man is exposed in the work place or in the environment and which might interact with one of the sensitive systems-the immunologic system of the body (Mehlman and Legator, 1991). The interaction of toxic chemicals with man or animal is dependent not only on chemical structure and exposure factors but also on age and nutritional status. This paper concludes that there is a need for creating awareness among the physicians, workers and regulatory personnel's about the role of the above factors in influencing the toxicity of chemicals (Cohen, 2004).

World over, the discipline of toxicology is rapidly advancing with multidisciplinary inputs. Both regulatory and investigative toxicology are being pursued in R&D laboratories, medical colleges and drug houses (Collins *et al.*, 2008). Regulatory toxicology is being predominantly pursued in drug houses. Adequate facilities for regulatory toxicology are available in the country and investigators are aware of Good Laboratory Practices (GLP) and observe them. Under investigative toxicology, current stress is on understanding the toxic effects of chemicals at molecular level to predict the toxicogenic potential of chemicals and identify the specific targets of toxicity (Conolly, 2009). The US Environmental Protection Agency is responsible for establishing criteria and standards to control potentially harmful chemicals entering the environment. The determination of whether a chemical is harmful is obtained from toxicity dose response tests using various organisms, including vertebrate, invertebrate, aquatic, and terrestrial organisms. The toxicants can be transferred through different pathways like in environment, health etc (Pesch *et al.*, 2004).

1) Toxicological Approach to Environment

Environmental pollution is a major natural hazard, and there are numerous reports of air pollution related wide spread sickness and in many cases, death in large numbers of domestic animals (Dogra, 1993). Substances responsible for environmental pollution may exist in diverse nature and physical form. These may be dusts, fumes, smokes or mists. The ecosystem around highways and other roads is subject to contamination by vehicular exhaust containing many toxic materials (Ward, 1990). A major contaminant is lead from alkyl lead compounds that have long been used as antiknock additives to petrol. Toxicity of lead in roadside environment is probably also influenced by contamination from other materials, including such heavy metals as cadmium and zinc (Saxena, 1993). The ambient air and water quality in which the man lives can significantly effect the toxicity of the chemicals and hence can play an important role in the risk assessment (Snyder and Green, 2001). The socioeconomic and biological pollution factors such as decay, refuse, burning, sewage and solid waste accumulation, epidemic pathogen, undesirable weeds, significantly alter the air and water quality an add to toxicological problems in developing countries (Chandra, 1979).

a) Environmental Risks

i) Air Pollution

The role of automobile exhaust as the major source of gaseous pollutant consisting of carbon monooxide, hydrocarbons, aldehydes, alcohols, nitrogen oxides and particulate matter like lead is significant. A recent survey around Calcutta area estimated that a vehicular exhaust fumes alone comprises more than 20 per cent of the daily pollutants (Tsai *et al.*, 1992). The major pollutants consisted of CO 177 tons, sulfur dioxide, oxides of nitrogen and poly nuclear aromatic hydrocarbon 63 tons each. The clinical picture of tetraethyl lead toxicity is one of the disturbed sleeps, nervousness followed by behavioral change confusion and abrupt onset of mania or convulsion. US Environmental Protection Agency as suggested measure to reduce pollution by use of lead free gasoline, return of low compression ratio engines and other likes wise steps. The simplest and least costly way to offset the octane number lowering effect of reduced lead is to increasing the proportion of aromatic hydrocarbons (Thompson *et al.*, 2008). Increasing use of aromatic hydrocarbons might be more dangerous than lead compounds, which they replace. The concentration of polynuclear hydrocarbon is much higher in the diesel engines exhaust than that of petrol one. Incidence of cancer have been reported from workers working with cutting oils. Aromatic hydrocarbons are primarily narcotic and have particular affinity for nerve tissue because of its high lipid content. Other hazard is the tendency to cause the dermatitis. Toulene and xylene also have irritating effect (Benigni and Bossa, 2008).

ii) Water Pollution

The wide scale production, transport, use and disposal of petroleum globally, it has become a leading contaminant in prevalence and quantity in many environments, but especially the ocean (Chemical Pollution; 1992). All kind of hydrocarbons processing industry constitutes a major potential for inland and coastal water pollution. Oilspills from tanker may become threat to aquatic and land life. The maximum permissible limit for oil and grease to be discharged to the inland water is 10 ppm (Bos *et al.*, 2004). The limit of oil discharge by ships in free zone area 50 miles away from coastal lines is 60 litres per mile travel, within overriding limit of 100ppm. The Marine Pollution Sub Committee (IMCS) has proposed 50 miles coastal zone in which deliberate discharge of oil is banned. Doubts have been expressed whether this give adequate protection for fisheries, marine wild life and tourism and to other publics hazards with the increasing use offshore oil precautions are necessary so that oil spill accidents like the one that occurred in 1969 at Santa Barbara Sea oil well (USA) can be avoided (ECHA, 2008).

iii) Occupational Risks

Workers in practically all occupations are exposed to petroleum hydrocarbons. Solvent in general neurotic toxic hydrocarbons particularly pose a hazard to the central nervous system. They can damage the brain and cause incurable nervous disorders. Some petroleum derivatives can cause liver and kidney injuries like carbon tetra chloride, chlorinated hydrocarbon benzene causes leukemia, some chlorinated hydrocarbons trichloroethylene, perchloroethylene, PAN are suspected to cause cancer. Emission of PAN by automobile exhaust is the main cause of urban environmental pollution (NIOSH; 1997). Methyl chloroform, TCE, toulene and carbon di sulfide and gasoline can contribute to cause a form of heart attack (arrhythmia). To prevent occupation hazard it is necessary that workers should not be exposed beyond TLV of chemicals and is vigilant on the hazard and should take protective steps for safe use of petrochemicals (Munro *et al.*, 2008).

iv) Risk to Livestock

Environmental pollution is a major human health hazard, but with the increasing industrial activity and spread to rural areas, livestock population has also become its prey. There are numerous reports of air pollutant related wide spread sickness and, in many cases, death in large number of domestic animals (Lillie, 1970). Before the advent of automation, animals like ponies, mules and donkeys were engaged in transport in mineral matter inside the mine or outside on the ground. Studies on pit ponies, which use to work underground in a coal mines for hauling the minerals, revealed accumulation of black pigment in the lungs. In some ponies small compact dust foci were observed in relation to the arterioles together with generalized emphysema (Burgoon and Zacharewski, 2008). There was formation of reticulin fibers among the dust cells but the fibrosis were moderate in comparison to that found in man. Deposit of anthracotic pigments and pneumoconiosis have been reported in fowl due to inhalation of soot near industries we have undertaken survey to study the effect of environmental pollution by coal mining and its transport on livestock population, residing in mining areas of Bihar, West Bengal, Orissa, Andhra Pradesh and Karnatka. As many as 7584 animals slaughtered for food purposes were examined postmortem. Body organs were collected from 596 animals for histopathological examination and particulate load studies, as their lungs showed varying sized mottling with black pigment (Scanlon, 1979).

v) Risk to Genetic Factors

It has been well known that response of the human beings to certain drugs and chemicals is variable due to genetic variability. The correlation between Glucose-6-phosphate dehyrogenase deficiency and hemolytic anemia and one antitrypsin deficiency and pulmonary emphysema is well known. Sickle cell trait, red blood cell abnormalities affect the CO and CN poisoning (Dogra R.K.S; 1979). Recently genetic polymorphism in drug metabolism has assumed a great significance in the consideration of risk due to xenobiotics. The genetic variation in rhodanase and paraoxanase activities significantly influences the toxicity of CN and parathion (Caporaso, 2007).

2) Toxicological Approach to Health Risk Assessment

The toxicology of health risk assessment can be summarized under the following sub headings:

A) Occupational Health Risk Assessment

The risk of committing of some illness of acute or chronic nature may be associated with practicing some trade or occupations. However, occupational health risk can be identified qualitatively and quantitatively by an occupational health expert looking after the health aspect of the involved workers (Hunter, 1974). Apart from direct assessment or measurement of chemical or physical factors in the work environment, the medical surveillance of workers engaged in a particular occupation constitutes a very relevant and a useful albeit indirect indicator of the prevailing environmental hygienic conditions. This medical surveillance can be of two types:

(a) Pre-employment and periodical medical examination of the shop floor workers (Axelrad *et al.*, 2007), and/or

(b) Estimation of biological limit values (BLVs) for monitoring the worker exposure other than reliance on the TLVs for industrial air (Schilling, 1973).

(a) Pre and Periodical Medical Examination

1) Pre Employment Medical Examination

This duty should performed by the occupational health specialized medical specialist. These should include some very basic and relevant clinicopathological examination such as recording of blood pressure, acquity of vision, stool, urine examination, sputum, X ray chest, ECG recording etc. This whole case sheet will constitute his baseline data. The purpose of pre placement examinations is to place the right man in the right job (Crump *et al.*, 2010).

2) Periodical Medical Examination

Subsequently during periodical medical examination, apart from thorough clinical checkup, only relevant parameters likely to be affected may be given more attention. The periodicity of such clinical checkups can be decided by the industrial physician or individual industry basis (Daston, 2008).

(b) Biological Limit Values (BLVs)

These constitute special tests designed specifically to monitor the worker exposure. The biologic measurement on which the BLVs are based can furnish two kinds of information useful in the control of worker exposure (Dix *et al.*, 2006)

1. Measure of the worker overall exposure.
2. Measure the workers individual and characteristic response (Proctor and Hughes; 1978). Some special test may also be included only in specialized industries *e.g.*, blood sugar and blood insulin estimation, blood hormones levels etc. in relevant drug or chemical industry dealing with such drugs or chemicals which may influence the parameters of the employee, and serum or RBC cholinesterase levels in the pesticides industry where OP or carbamate pesticide are being handled. The biologic limits (BLVs) may be used as an adjunct to the TLVs for air, or in place of them. In a large number of situations where TLVs are not available environmental monitoring facility does not exist, the BLVs can be extremely useful indicators (Euling *et al.*, 2010).

Risk Assessment in Terms of Standardized Morbidity or Mortality Ratio (SMR)

The estimate of persons falling ill or dying in particular trade in a country belonging to various age groups and of a particular sex can be compared with national morbidity or mortality rates per specified period of time multiplied by hundred gives the figure (Clayton G.D Clayton F.E; 1981)

$$\text{SMR formula} = \text{observed death (sickness) rate}^a / \text{expected death (sickness)rate}^a \times 100$$

where,

a: For the same sex and same age group

B) Occupational Health Risk Assessment due to Petroleum Hydrocarbons

Petrol has become an essential commodity in our daily life. Intoxication by gasoline and related products can lead to serious ill health. Chronic occupational intoxication has been studied by exposing human volunteers to air containing gasoline vapors. Possible health hazards have also been highlighted by international labour organization. The survey comprises recording of occupational history, clinical examination, and biochemical tests in blood and urine and psychological investigation (Divine B.G, Borowl V., Kaplon J.D; 1985). Prominent symptoms in exposed workers were headache, heaviness of head and lacrimation. Most common signs are prominent bulber vessels, coated and furred tongue and congested throat. Many control strategies are being developed with respect to the environmental impact assessment to combat oil pollution due to petroleum hydrocarbons including improved legislation, comprehensive environmental assessment and management plans, better prevention and cleanup technologies, better design and executed monitoring programs and research on a number of key uncertainties associated with oil pollution and risk assessment (Feero *et al.*, 2008).

C) Health Risk Assessment due to Combined Effects of Pathogens and Pollutants

Attempts have been made to focus on various industrial pollutants to describe the potential toxic effects of pollutants rendering the host more susceptible to opportunistic infections and resulting in increased morbidity and mortality (Hall J.G; 1987). In recent years the role of infection the augmentation of occupational respiratory diseases due to inhalation of particulate, fibrous dust, gases and aerosols have been recognized. The lung diseases, pulmonary lesions, massive fibrosis are some of the diseases which are caused due to the exposure to asbestos, silica, heavy metallic mineral dusts, coal mine dust. Since the oral route is another most important modes of entry for the long term low level exposure to a variety of chemical pollutants much more needs to be known about likely alteration in the immune homeostasis of gastrointestinal tract as well as the functioning of the other systems of the body (Georgopoulos *et al.*, 2009). Because of the serious health implication of immune toxicity caused by pollutants attention should be given to the ways of promoting of a greater awareness among clinicians and occupational physicians of the possibility that pollutants of many different types along with infection can produce severe immune alteration which may have undesirable health consequences

D) Environmental and Occupational Health Risk assessment Due to Heavy Metals

Metals are present every where in the environment in their appropriate level but this level exceeds beyond limit, their exposure poses a danger to human health. Burning of coal, smoke of domestic chimney and motor vehicle exhaust are also responsible for metal emission. When these metals reach the river water, they pollute them. The fishes, birds and even mammals depending on them take-up these toxic metals through this food chain general public is ultimately affected (Friberg and

Nordberg, 1986). Effluents contains mercury are also being discharged into waterways in our country by industries. Fishes are affected found in water system near the towns having chloralkali and paper making industries. In acute and chronic toxicity of arsenic it has symptoms like cough, dyspnea, chest pain, dermatitis, nasal mucosol irritation, laryngitis, mild bronchitis and conjunctivitis, neuronal and cardiovasculatory disorders are reported. Cadmium poising in acute stage results in increased salivation, vomiting, abdominal pain, diarrhoea, and headache, but on inhalation general weakness, respiratory problem and shortness of breath (Krishnamurthy and Vishwanathan; 1991). Lead intoxication causes fatigue, disturbance of sleep and constipation in early stages, while more prolong and colic, anemia, and neuritis may follow severe exposure (Chiu *et al.*, 2010).

The assessment of the above said health effects can be monitored by the following methods:

1. Ambient Monitoring

Ambient monitoring assess the health risk by measuring the external exposure to the chemical that is concentration in air, food, water etc (Anderson and D'Appollonia; 1978).

2. Biological Monitoring of Exposure

Biological monitoring of exposure assesses the health risk through the evaluation of the internal dose. The assessment is normally carried out on an individual basis. The internal dose may be estimated directly by measuring the chemical or its metabolites in biological media or indirectly by quantifying nonadverse biological effects related to internal dose (Cairns;1982).

3. Health Surveillance or Biological Monitoring of Effects

Health surveillance must be clearly distinguished from ambient and biological monitoring of exposure. Later attempts to detect unhealthy exposure conditions (health risk), health surveillance evaluates the health status and aims at identifying individuals with signs of adverse effects (Holdgate;1979).

4. Ecological Monitoring

Wilkinson (1983) considered ecological monitoring to be the assessment of effects of toxicants and pollutants either by measuring their accumulation in the organisms and other than humans or by looking for abnormal ecological effects at the level of the species, community or ecosystem. Eco-monitoring and chemical monitoring according to Wilkinson (1983) should go hand in hand. Unlike the later the former does not need frequent sampling and can also indicate the past exposure. In Eco-monitoring bioaccumulation, bio-organisms or laboratory bioassays are used along with ecological population dynamics. Connell and Miller (1984) considered ecological monitoring to have two facts:

1. Factor monitoring which is pollutant measuring in different parts of the environment.
2. Target monitoring which is quantification of effects on species of ecosystem (Margalef, 1985).

Testing of toxicity

The multifold purposes of toxicity testing includes (Gant, 2007):

1. Identification of potential victim and causative factor
2. Environmental fates and response
3. Evaluation of safety
4. Containment of short term and spillage effect, and
5. Prevention of long term effects Pasthumus (1985).

I. Quantitative Approach of Risk Assessment of Human Health

Risk assessment provides information about the consequences of possible action to the management. Important decision that could use risk estimates includes selecting waste treatment or disposal method, remediating contaminating sites, minimizing waste generation, siting new facilities and developing new products (Gaylor and Kodell, 2000). Risk is classically defined as the probability of suffering harm or loss. However, the actual definition depends on whether the resulting harm can be measured or not. When the resulting harm is measurable the risk can be calculated as (Gohlke *et al.*, 2009),

$$Risk = probability\ of\ action \times severity\ of\ consequence$$

In a number of cases if it is not possible to quantify the severity of consequence for *e.g.* Death cannot be quantified; no one is ever slightly dead. For such consequences (Kavlok *et al.*, 2009),

$$Risk = probability\ of\ the\ harm\ occurring$$

II. Risk Assessment Methodology

Risk assessment requires a clear understanding of what chemicals are present at a site, their concentration and special distribution and how they could move in the environment from the site to potential receptor point. It is a four-step process (Lin *et al.*, 2006).

a) Hazard Identification

A large number of chemicals may be present at waste site, however all of them may not be important in terms of posing a significant risk. There are few steps in identifying the surrogate chemicals (NRC, 2007a).

1. Initial Screening

☆ The chemicals are first sorted out by medium of transport for both carcinogens and non-carcinogens.

☆ For each deducted chemical the mean and range of concentration value at the "source" is tabulated.

☆ The toxicity scores (TS) for each chemical in their respective media are determined (Gwinn *et al.*, 2009) :

Non Carcinogens

$$TS = C_{max}/RfD$$

where,

C_{max}: maximum concentration

C_{max}: maximum concentration

RfD: chronic reference dose

Carcinogens

$$TS= SF \times C_{max}$$

where,

SF: Slope factor (also called carcinogen potency factor)

☆ For each exposure route, the chemicals are ranked in terms of toxicity scores

☆ For each exposure route, those chemical comprising 99 per cent of the total score are the " surrogate" chemicals

2. Final Screening

In addition to the toxicity other factors need to be considered for the selection of the surrogate chemicals. Some of these factors are (Khoury *et al.*, 2009a) :

☆ Concentration range or mean concentration

☆ Frequency of detection

☆ Mobility

☆ Persistence in the environment

☆ Treatiblity

b) Exposure Assessment

The objective of exposure assessment is to estimate of doses of surrogate chemicals to which the receptors are exposed. Exposure assessment consist of (Mohrenweiser, 2004):

☆ Identifying the source of contamination

☆ Estimating the spatial distribution of surrogate chemicals at the source

☆ Identifying the mechanism by which the chemicals are released

☆ Determining how surrogate chemicals are transported to current and potential receptor points

☆ Identifying the receptor points

☆ Determine the relevant exposure scenarios

☆ Estimating exposure point concentration

☆ Estimating both short and long term exposure in terms of doses by different exposure routes

c) Toxicity Assessment

In toxicity assessment a slope factors and references doses for surrogate chemicals are determined for each exposure route. In addition, the uncertainty inherent in the numerical values SF and RfD and the effect of uncertainty on the estimates of risk should also be analyzed. Slope factor is calculated as the 95 per cent upper confidence limit of the dose response curve and is expressed as the inverse of dose (kg day/mg). Reference dose can be defined as, unlike carcinogen, non-carcinogens do exhibit a threshold level. The uncertainties in values of slope factor and RfD's can be determined the toxic effect of chemical on testing of experimental animals, particularly small mammals, because of the lack of sufficient human data (Naciff *et al.*, 2009).

d) Risk Characterization

In this final stage of risk assessment, the magnitude of both carcinogenic and noncarcinogic is calculated. Risk calculation is performed for the relevant exposure scenarios and the three step exposure routes. Risk estimates based on average concentration is better for chronic exposure while "maximum concentration" base and calculation provide a useful upper bound estimate of potential risk. Also, the formula is given below pertaining to incremental risk (Ovacik *et al.*, 2010).

Carcinogic Risk

$$CR = CDI \acute{} SF$$

where,

    CR: Carcinogenic risk

    CDI: Chronic daily intake (mg/kg day)

    SF: Carcinogenic slope factor (kg day/mg)

Non-Carcinogic Risk

$$HI = CDI/RfD$$

where,

    HI: Hazard index (dimensionless)

    CDI: Chronic daily intake (mg/kg day)

    RfD: Reference dose (mg/kg day)

Clearly HI less than 1 is desirable.

Assessing Risks from Health Hazard: An Imperfect Science

Risk assessment had become so entangled with politics that many public observers felt that EPA was acting as an advocate for the very industries it was suppose to regulate. For *e.g.* EPA had concluded that there was no significant effect on health risk to workers from exposure to formaldehyde, a chemical used to make particle board, plywood and some permanent press fabrics. In an effort to make risk assessment more impartial, Ruckelshaws argued that scientist should first make an " objective" study of the extent of the risk from exposure to a particular hazardous chemical or situation. Ruckelshaws intend was to create a special authority and credibility for risk assessment. He wanted to build a strong scientific foundation upon which EPA and society at large could balance social, economic and political concerns and reach sensible decision about managing environmental risks (USEPA, 2009).

Evaluating and environmental hazard to see how many people might be at risk is difficult because analyst must follow that hazard through whatever twist and turns it takes in the real world. Analyst should determine how the potential threats are released into, and move through the environment. They have to figure out how much of the substance people might eat, breath, or otherwise take-up and then estimate how much of it they would absorb. Finally, analyst must determine just how much of a hazard the absorbed level of the substance poses.

Conclusion

It is apparent from the above that problems of toxicology in developing countries are much more diverse and demand a different approach for the risk assessment and management. There is a need for creating awareness among the physicians, workers and regulatory personnel's about the role of the above the factor in influencing the toxicity of chemicals. The role of biological monitoring and Eco-toxicology in identification of problems and their prediction and prevention connected with fresh water pollution is clear although the science is still in infancy and needs considerable advancement. Bio-monitoring studies can also be applied for testing the efficacy of effluent treatment procedure and for comparing relative pollution loads of different habitats. Thus, health risk assessment with respect to toxicology proves to be a tool in the branch of toxicology.

References

Ali M. (1993). *Principles of toxicology*. Pub: I.T.R.C pp. 1-7.

Anderson, P.D and D'Appollonia, S. (1978). *Principles of ecotoxicology*. G.C Buttler (ed) Scope 12. ISCU-SCOPE, Pub: John Wiley and sons; Chichester; pp.187-221.

Anonymous, (1981). *Patty's Industrial hygiene and Toxicology*. Clayton J.D and Clayton F.E, Edits. Pub: John Wiley and sons, New York, U.S.A

Anonymous, (1992). Chemical pollution (A global review) Earth watch; U.N.E.P., Geneva; pp.1-4.

Anymous (1978). *Chemical Hazards of the work place*. Proctor N.H and Hughes J.P, Edits. Pub: Lippincott Company, Philadelphia, Toronto, U.S.A

Axelrad, D.A., Bellinger, D.C., Ryan, L.M., Woodruff, T.J. (2007). *Dose–response relationship of prenatal mercury exposure and IQ: an integrative analysis of epidemiologic data*; In: *Environ. Health Perspect.* 115 (4) pp. 609–615.

Benigni, R. and Bossa, C. (2008). *Structure alerts for carcinogenicity, and the Salmonella assay system: a novel insight through the chemical relational databases technology*. In: Mutat. Res. 659, pp. 248–261.

Bos, P.M.J., Baars, B.J. and van Raaij, M.T.M. (2004). *Risk assessment of peak exposure to genotoxic carcinogens: a pragmatic approach.* In: Toxicol. Lett. 151 (1); pp. 43–50.

Browing, E. (1987) *Toxicity of metabolism of industrial solvent.* Pub: Elsevier Amsterdam. p. 1.

Burgoon, L.D. and Zacharewski, T.R. (2008). *Automated quantitative dose–response modeling* and point of departure determination for large toxicogenomic and high-throughput *screening data sets. Toxicol. Sci.* 104 (2); pp. 412–418.

Cairns, J. (1982); Biological monitoring of water pollution; Pub: Pergamon Press.

Caporaso, N.E. (2007). *Integrative study designs – next step in the evolution of molecular epidemiology?* In: Cancer Epidemiol. Biomarkers Pre 16; pp. 365–366.

Chandra, R.K. (1979). Bulletin of the World Health Organization. pp.57, 167.

Chiu A. Weihsueh, Euling Y. Susan, Scott Siegel Cheryl and Subramaniam P. Ravi (2010) *Approaches to advances quantitative human health risk assessment of environmental chemicals in the post genomic era;* In: Toxicology and Applied Pharmocology; Elsevier; pp. 1-15.

Cohen, S.M. (2004). *Risk assessment in the genomic era;* In: Toxicol. Pathol. 32 (Suppl 1), pp. 3–8.

Collins, F.S., Gray, G.M., Bucher, J.R. (2008). *Toxicology. Transforming environmental health protection;* In: Science 319 (5865); pp 906–907.

Conolly, R.B. (2009). *Commentary on "Toxicity testing in the 21st century: implications for human health risk assessment" by Krewski et al;* In: *Risk Anal.* 29 (4); pp.480–481.

Crump, K.S., Chen, C., Chiu, W.A., Louis, T.A., Portier, C.J., Subramaniam, R.P. and White, P.D. (2010). *What role for biologically-based dose–response models in estimating low dose risk?* In: Environ. Health Perspect. Epub 2010. 4 Jan. doi:10.1289/ehp.0901249.(available at http://dx.doi.org/).

Daston, G.P. (2008). *Gene expression, dose–response, and phenotypic anchoring: applications for toxicogenomics in risk assessment.* In: Toxicol. Sci. 105 (2); pp. 233–234 Epub 2008 Aug 6.

Divine, B.J. Brawn, V. Kaplan, J.D. (1985) *The Texaco Mortality Study – Mortality among refinery petrochemical and research worker.* In: J.Occupation. Med. pp.27, 445.

Dogra, R.K.S. (1979a). *Environmental pollution mediated risk to live stock.* Pub: I.T.R.C. pp1-9.

Dogra, R.K.S. (1979). *Pathological studies of pulmonary tissue and regional lymph nodes of domesticated animals in various mining areas of India.*Ph.D. Thesis, Agra University, Agra;

Dix, D.J., Gallagher, K., Benson, W.H., Groskinsky, B.L., McClintock, J.T., Dearfield, K.L. and Farland, W.H. (2006). *A framework for the use of genomics data at the EPA.* Nat.Biotechnol. 24 (9); pp. 1108–1111.

ECHA, (2008). *European Chemicals Agency. Guidance on information requirements and chemical safety assessment.* In: Chapter R.8: Characterisation of dose [concentration]-response for human health. May 2008.

Euling, S.Y., Makris, S., White, L., Chiu, W., Benson, R., Thompson, C., Keshava, N.,Keshava, C., Ovacik, A.M., Sen, B., Gaido, K., Androulakis, I.P., Hester, S., Wilson, V., Gray Jr., L.E., Foster, P. and Kim, A.S. (2010). This issue. *An approach to using toxicogenomic data in risk assessment: the dibutyl phthalate case study.* In: Toxicology and Applied Pharmacology Toxicogenomics in Risk Assessment Special Issue. Volume, Issue, Page Numbers TBD.

Feero, W.G., Guttmacher, A.E., and Collins, F.S. (2008). *The genome gets personal – almost.* JAMA 299 (11); pp. 1351–1352.

Friberg, L. Nordberg, G.F and Vouk, V.B, (1986). *Handbook on the toxicology of metals Part I and II,* Pub: Levis Publishers, Michigan.

Gant, T.W. (2007). *Novel and future applications of microarrays in toxicological research.* In: Expert Opin. *Drug Metab. Toxicol.* 3 (4); pp. 599–608.

Gaylor, D.W. and Kodell, R.L. (2000). *Percentiles of the product of uncertainty factors for establishing probabilistic risk doses;* In: *Risk Anal.* 20; pp. 245–250.

Georgopoulos, P.G., Sasso, A.F., Isukapalli, S.S., Lioy, P.J., Vallero, D.A., Okino, M. and Reiter, L. (2009). *Reconstructing population exposures to environmental chemicals from biomarkers: challenges and opportunities.* In: *J. Expo. Sci. Environ.*Epidemiol. 19 (2); pp 49–71.

Gohlke, J.M., Thomas, R., Zhang, Y., Rosenstein, M.C., Davis, A.P., Murphy, C., Becker, K.G., Mattingly, C.J. and Portier, C.J. (2009). *Genetic and environmental pathways to complex diseases.* In: BMC Syst. Biol. 3; pp 46.

Gwinn, M., Guessous, I. and Khoury, M.J. (2009). *Invited commentary: genes, environment, and hybrid vigor.* In: Am. J. Epidemiol. 170; pp 703–707.

Hall, J.G. (1987). *In Immunotoxicology.* (A.Berlin, J.Dean, M.H Draper, E.M.B Smith and F.Spreafico Edits.), Pub: Martin Nijhoff Publishers, Dordracht/Boston/Lan Caster,

Hattis, D.and Kennedy, D. (2000). *Assessing Risk from health hazard: An Imperfect Science.* Massachusetts institute of technology and Harvard University Cambridge; IIT Kanpur; pp.77-84.

Holdgate, M.W (1979). *A perspective of Environmental pollution.* Pub: Cambridge University Press; Cambridge.

Hunter, D. (1974) The diseases of occupation; 5th edition; Pub: English University Press Limited. London; U.K.

Kavlock, R.J., Austin, C.P. and Tice, R.R. (2009). *Toxicity testing in the 21st century: implications for human health risk assessment. In:* Risk Anal. 29 (4); pp 485–487

Khoury, M.J., Feero, W.G., Reyes, M., Citrin, T., Freedman, A. and Leonard, D., the GAPPNet Planning Group. (2009a). The genomic applications in practice and prevention network. Genet. Med. 11; pp 488–494.

Krishnamurthy, C.R. and Vishwanathan P. (1991). *Toxic metals in the Indian environment.* Pub: Tata Mc Graw Hill Pub.co.Ltd. New Delhi.

Lillie R.J. (1970) *Air pollution effecting the performance of domestic animals- A literature review.* In: Agriculture Handbook number 380; Agri.Res.U.S.D.A. Washington.

Lin, B.K., Clyne, M., Walsh, M., Gomez, O., Yu, W., Gwinn, M. and Khoury, M.J. (2006). Tracking the epidemiology of human genes in the literature: the HuGE Published Literature database. In: *Am. J. Epidemiol.* 164 (1); pp. 1–4.

Margalef, R. (1985). *Pollutants and their Ecotoxicological significance.* H.W Nurnberg (Ed); Pub: John Wiley and Sons; pp. 149-175.

Mehlman, M.A and Legator, L.S. (1991). *Dangerous and cancer causing properties of products and chemicals in the refining and petrochemical industry.* In: Toxicology Industrial Health. pp. 7,207.

Mohrenweiser, H.W. (2004). *Genetic variation and exposure related risk estimation: will toxicology enter a new era? DNA repair and cancer as a paradigm.* In: *Toxicol. Pathol.* 32 (Suppl 1); pp 136–145.

Munro, I.C., Renwick, A.G. and Danielwska-Nikiel, B. (2008). *The threshold of toxicological concern (TTC) in risk assessment.* In: Toxicol. Lett. 180; pp 151–156.

Naciff, J.M., Khambatta, Z.S., Thomason, R.G., Carr, G.J., Tiesman, J.P., Singleton, D.W., Khan, S.A. and Daston, G.P.(2009). *The genomic response of a human uterine endometrial adenocarcinoma cell line to 17alpha-ethynyl estradiol.* In: Toxicol. Sci.107 (1); pp. 40–55.

Niosh, (1977). *Occupational exposure of refined petroleum solvents.* Pub: US departmental health educational welfare, pp.1.

NRC (National Research Council). (2007a). *Applications of Toxicogenomic Technologies to Predictive Toxicology and Risk Assessment.* In: National Academy Press, Washington, DC.

Ovacik, A.M., Ierapetritou, M.G., Sen, B., Gaido, K. and Androulakis, I.P. (2010). This issue. A pathway activity method applied to microarray studies of the testis following in utero exposure to dibutyl phthalate (DBP). In: Toxicology and Applied Pharmacology Toxicogenomics in Risk Assessment Special Issue. Volume, Issue, Page Numbers TBD.

Pandya, K.P. (1993). *Environmental and Occupational health risk due to petroleum hydrocarbons.* Pub: I.T.R.C. pp. 1-9.

Pesch, B., Brüning, T., Frentzel-Beyme, R., Johnen, G., Harth, V., Hoffmann, W., Ko, Y., Ranft, U., Traugott, U.G., Thier, R., Taeger, D. and Bolt, H.M. (2004). *Challenges to environmental toxicology and epidemiology: where do we stand and which way do we go?* Toxicol. Lett. 151 (1); pp. 255–266.

Posthumus, A.C. (1985). *Pollutants and their Ecotoxicological Significance.* H.W. Nurnberg (Ed) Pub: John Wiley and Sons; pp. 55-65.

Saxena, D.K. (1993). *Environmental and occupational health risks due to heavy metals.* Pub: I.T.R.C. pp.1-9.

Scanlon, P.F. (1979). *Lead contamination of mammals and invertebrates near highways with different traffic volumes.* In: Animals as monitors of environmental pollutants. Pub: The National Research Council; Washington, pp. 200-208.

Schilling, R.S.F (1973). *Occupational Health Practice,* 1st edition, Butterworth and corporation Publishers Ltd. London, U.K.

Shekhar, R. (1999) *Quantitative Risk Assessment of human Health.* Vol. I; Pub: IIT Kanpur; pp.46-62.

Snyder, R.D. and Green, J.W. (2001). *A review of the genotoxicity of marketed pharmaceuticals.* In: Mutat. Res. 488 (2); pp 151–169.

Thompson, C.M., Sonawane, B., Barton, H.A., DeWoskin, R.S., Lipscomb, J.C., Schlosser, P., Chiu, W.A. and Krishnan, K. (2008). *Approaches for applications of physiologically based pharmacokinetic models in risk assessment.* In: J. Toxicol. Environ. Health B Crit. Rev. 11 (7); pp. 519–547.

Tsai, S.P. Dowd, C.N, Cowles S.R and Ross C.E. (1992). A Prospective Study of morbidity pattern in a petroleum refinery and chemical plant. In: Br. J. Ind. Med. pp. 49,576.

U.S. EPA (2009). *The First International Workshop on Virtual Tissues: v-Tissues 2009.*

http://www.epa.gov/ncct/virtual_tissues/.

Ward N.I. (1990). *Multimetal contamination of British motorway environment;* In: Science of total Environment. pp.93, 393-401.

16

Health Impacts of Natural Disasters: Reference to Hydro-meteorological Disaster Management

J Kishore, Ravneet Kaur and Anil K. Gupta

Definition and Health Impact of Disaster

Disaster is any occurrence that causes damage, ecological disruption, loss of human life and deterioration of health and health services on a scale sufficient to warrant an extraordinary response from outside the affected community[1]. The American Red Cross defines a disaster as an occurrence such as hurricane, tornado, storm, flood, high water, wind driven water, tidal wave, earthquake, drought, blizzard, pestilence, famine, fire, explosion, volcanic eruption, building collapse, transportation wreck, or other situation that causes human suffering or creates human needs that the victims cannot alleviate without assistance[2]. From these definitions it is evident that health impact and disasters are inseparable.

There has been an increase in the number of natural disasters over the past years. Due to increase in urbanization and population growth, natural disasters lead to greater losses, as a result of which their impact is now felt to a larger extent. In 2008, 321 natural disasters killed 235 816– a death toll that was almost four times higher than the average annual total for the seven previous years. Asia, the worst-affected continent, was home to nine of the world's top 10 countries for disaster-related deaths. Between the years 1991 to 2000 Asia has accounted for 83 per cent of the population

Table 16.1: Health (related) effects of natural hazards.

Health (Related) Effects	Earthquake	Floods	Land-slides	Epidemics	Conflict Situation
Deaths/Severe injuries	Many	Few	Many	Many	Many
Requiring extensive treatment	Many	Few	Few	Few	Many
Increased risk of epidemics	Yes	Yes	Yes	–	Yes
Damage to water systems	Severe	Light	Severe (but localized)	None	Limited (depends on the factions fighting)
Damage to health facilities	Severe (structural and equipment)	Severe (equipment usually)	Severe (but localized)	None	Limited (depends on the factions fighting)
Damage to health services	High	High	Low	Moderate	High
Food shortage	Possible (due to distribution problems)	Common	Common (but localized)	None	Common (in prolonged conflicts)
Major population movements	Common (generally limited)	Common	Common (generally limited)	Rare	Common (generally limited)

Table 16.2: Outbreaks following natural disasters in India.

Year	Type	Place	Death/ (Injuries)	Outbreak
1999	SuperCyclone	Orissa	9,500	Leptospirosis
2001	Earthquake	Bhuj,Gujarat	19,800 (1.66 lakhs)	Sporadic incidence of water borne diseases
2004	Flood	Gujarat, Assam, Bihar	N.A.	Sporadic incidence of diarrhoeal diseases
2004	Tsunami	A&N Islands, T.N., A.P., Kerala, Pondicherry	10,749 (5,640 missing)	Focal outbreak of measles in coastal Tamil NaduIncreased incidence of malaria in endemic areas of southern group of A&N I
2005	Flood	Mumbai, Maharashtra	262	Leptospirosis (366 deaths)

affected by disasters globally. Within Asia, 24 per cent of deaths due to disasters occur in India. Many parts of the Indian sub-continent are susceptible to different types of disasters owing to the unique topographic and climatic characteristics. For example,

☆ Coastal States particularly in the East Coast and Gujarat are vulnerable to cyclones.

☆ 4 crore hectare land mass is vulnerable to floods.

☆ 68 per cent of net sown area is vulnerable to drought.

☆ 55 per cent of total area is in Seismic Zones III - V, and vulnerable to earthquakes.

☆ Sub-Himalayan/Western Ghat is vulnerable to landslides[3].

Major issue raised in this article is the impact of disasters on the health and well being of people. It is witnessed that large number of people are killed and injured, displaced, and subjected to greater risk of epidemics in post disaster phase.

There are three types of Biological Disasters:

1. Post Disaster Epidemics which can complicate disaster and affect the relief activities;

2. Epidemics as the Disaster such as Severe Acquired Respiratory Syndrome, Bird Flue, Pandemic Flue, Plague, etc.

3. Biological Warfare due to deliberate release of microorganisms in the environment or attacks on the particular community.

The relative number of injuries and deaths differ depending on the type of disaster, density and distribution of population and degree of preparedness. Similarly damage to the health facilities also differ according to the type of disasters (Table 16.1).

The heightened risk of outbreaks of communicable diseases after a disaster can lead to widespread death and suffering. Frequently encountered post-disaster epidemic prone diseases (Table 16.2) include:

☆ Water borne diseases like Acute diarrheal diseases (including cholera, dysentery), enteric fever, acute jaundice syndrome

☆ Vector borne diseases *i.e.* malaria, dengue, acute encephalitis (including JE), leptospirosis

☆ Vaccine preventable: measles

☆ Others: meningitis, HIV/AIDS, hemorrhagic fevers

☆ Conditions owing to lack of sanitation, emergence of new breeding sites, disruption of vector control activities and overcrowding in shelters and camps

Besides this, there can be long-term effect on psychological and social milieu of the victims. Disruption of social life further disturb the equilibrium of epidemiological triad, *i.e.* agent, host, environment.

Disasters affect both the developing as well as developed countries however the vulnerable in these countries tends to suffer more. Although only 11 per cent of the people exposed to natural hazards live in developing countries, they account for more than 53 per cent of global deaths due to natural disasters. The differences in impact suggest there is great potential to reduce the human death toll caused by disasters in developing countries. One of the reasons for huge suffering and deaths is human inaction during and after disaster.

What is disturbing is the knowledge that these trends of destruction and devastation are on the rise instead of being kept in check.

Economic Loss

Disasters also exact a devastating economic toll. In 2008, disasters cost an estimated US$ 181 billion – more than twice the US$ 81 billion annual average for 2000–2007. The Sichuan earthquake was estimated to cost some US$ 85 billion in damages, and Hurricane Ike in the United States cost some US$ 30 billion[3].

These economic losses are both immediate as well as long term in nature and demand additional revenues. Also, as an immediate fall-out, disasters reduce revenues from the affected region due to lower levels of economic activity leading to loss of direct and indirect taxes. In addition, unplanned budgetary allocation to disaster recovery can hamper development interventions and lead to unmet developmental targets.

Disasters may also reduce availability of new investment, further constricting the growth of the region. Besides, additional pressures may be imposed on finances of the government through investments in relief and rehabilitation work

The dimensions of the damage emphasize the point that natural disasters cause major setbacks to development and it is the poorest and the weakest that are the most vulnerable to disasters. Given the high frequency with which one or the other part of the country suffers due to disasters, mitigating the impact of disasters must be an integral component of our development planning and be part of our poverty reduction strategy.

Disaster Management

Disaster Management can be defined as the organization and management of resources and responsibilities for dealing with all humanitarian aspects of emergencies, in particular preparedness, response and recovery in order to lessen the impact of disasters[1].

Phases in the Management of Disasters

There are four essential phases in the management of disasters:

1. Warning phase
2. Emergency phase
3. Relief phase
4. Recovery (Rehabilitation) phase

Warning Phase

Most of the climatic natural disasters can be predicted before hand. The health impact of emergencies and crises can be substantially reduced if both national and local authorities and communities in high-risk areas are well prepared and are able to reduce the level of their vulnerabilities and the health implications of their risks. Measures to be taken in this phase include awareness generation, improving community health security and relocation or protection of vulnerable populations or structures.

Disaster preparedness refers to measures taken to prepare for and reduce the effects of disasters. That is, to predict and, where possible, prevent disasters, mitigate their impact on vulnerable populations, and respond to and effectively cope with their consequences.

Emergency Phase

The emergency phase of a relief operation aims to provide life-saving assistance; shelter, water, food and basic health care are the immediate needs; along with a sense of humanity and a sign that someone cares. Subsequent needs include reconstruction and rehabilitation. These needs can continue for several years, particularly in the case of refugees and victims of socio-economic collapse. The first people to respond to a disaster are those living in the local community. They are the first to start rescue and relief operations. The focus should be on community-based disaster preparedness, which assists communities to reduce their vulnerability to disasters and strengthen their capacities to resist them.

Care of the Survivors

Injuries account for the vast majority of mortality experienced so far in countries affected by disasters. Falling structures, and waters full of swirling debris may inflicted crush injuries, fractures, and a variety of open and closed wounds. Appropriate medical and surgical treatment of these injuries is vital to improving survival, minimizing future functional impairment and disability and ensuring as full a return as possible to community life.

Care of injured survivors will depend on:

☆ Severity of injury

☆ The extent to which appropriate medical and surgical resources are available.

Many injuries are severe enough to lead to long term functional impairment and disability if not managed correctly e.g Spinal cord injuries, Crush injuries of extremities requiring amputation, Fractures that develop complications such as infection prioritized or alignment problems

Such injuries must be recognized and referred to specialty or tertiary hospitals that can manage such cases and are appropriately equipped with mobility aids, assistive devices and physiotherapy support.

Triage is an Important Consideration for Managing Injured Survivors

Patients should be categorized by severity of their injuries and treatment in terms of available resources and chances for survival. The underlying principle of triage is allocation of resources in a manner ensuring the greatest health benefit for the greatest number.

There is an internationally accepted colour code system. Red indicates high priority treatment or transfer, yellow signals medium priority whereas green indicates ambulatory patients and black for dead or moribund patients respectively.[4]

The initial assessment should be followed up with more detailed assessment during the rehabilitation and recovery phase as shown in the following flow diagram.

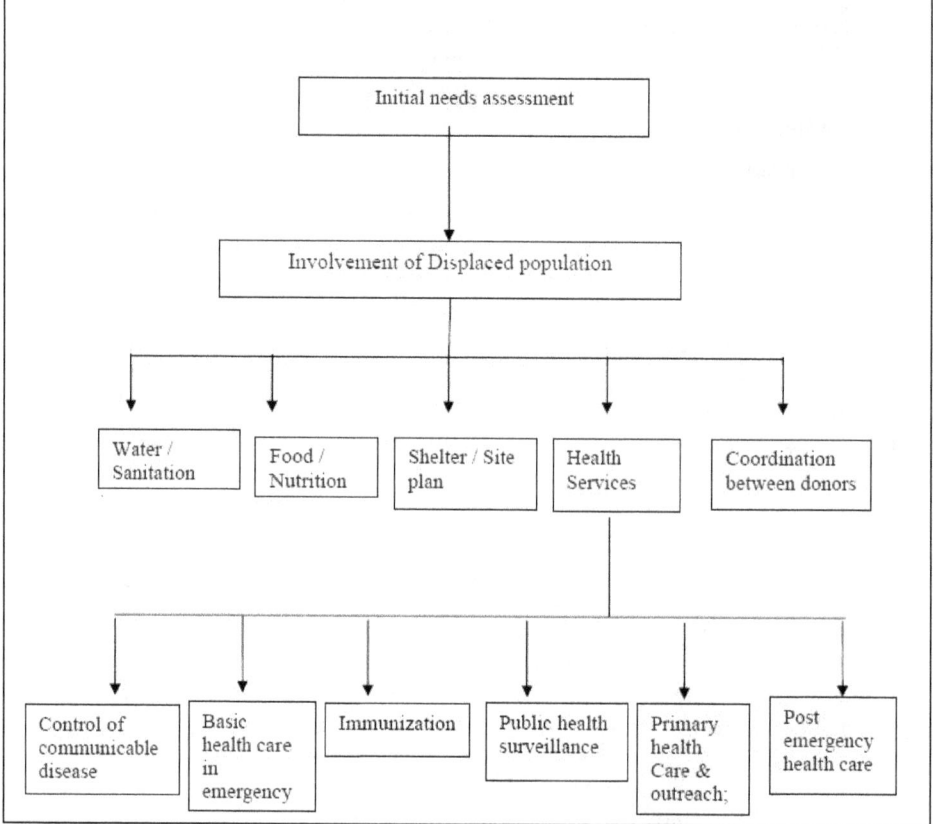

```
                    ┌─────────────────────────────┐
                    │   Initial needs assessment  │
                    └─────────────────────────────┘
                                   │
                                   ▼
              ┌──────────────────────────────────────────┐
              │   Involvement of Displaced population     │
              └──────────────────────────────────────────┘
                                   │
      ┌──────────┬──────────┬──────────┬──────────┬──────────┐
      ▼          ▼          ▼          ▼                     ▼
 ┌────────┐ ┌────────┐ ┌──────────┐ ┌────────┐      ┌──────────────┐
 │ Water /│ │ Food / │ │ Shelter /│ │ Health │      │ Coordination │
 │Sanitation│ │Nutrition│ │ Site plan│ │Services│      │between donors│
 └────────┘ └────────┘ └──────────┘ └────────┘      └──────────────┘
                                         │
   ┌──────────┬──────────┬──────────┬──────────┬──────────┐
   ▼          ▼          ▼          ▼          ▼          ▼
┌────────┐┌────────┐┌──────────┐┌──────────┐┌────────┐┌──────────┐
│Control │││Basic   ││Immuniz-  ││Public    ││Primary ││Post      │
│of      ││health  ││ation     ││health    ││health  ││emergency │
│communic││care in ││          ││surveill- ││Care &  ││health    │
│able    ││emerg-  ││          ││ance      ││outreach││care      │
│disease ││ency    ││          ││          ││;       ││          │
└────────┘└────────┘└──────────┘└──────────┘└────────┘└──────────┘
```

Care of the dead

Deaths due to disaster often require medico-legal investigation especially for identification. Dead bodies should be removed from the site of disaster and shifted to a mortuary. Identification of the bodies and dealing with the bereaved relatives is another important part of disaster management[4]

Relief Phase

This phase starts when aid from outside starts to reach the disaster area. In this phase, assessment is a crucial management task which contributes directly to effective decision-making, planning and control of the organized response. Assessment of needs and resources is required in all types of disasters, whatever the cause and whatever the speed of onset. Assessment is needed during all identifiable phases of a disaster: from the start of emergency life-saving through the period of stabilization and rehabilitation, and into long-term recovery, reconstruction and return to normalcy.

Three general priorities are to be identified for early assessments:

1. Location of problem
2. Magnitude of problem
3. Immediate priorities [2]

The initial assessment should be followed up with more detailed assessment during the rehabilitation and recovery phase as shown in the following flow diagram (Figure 16.1).

Camps may have to be organized for the displaced so that there is equity in distribution of resources and disease and disability are minimized. Support of the local administrative officers and community leaders is essential in this phase. Group leaders from the community can be trained for wound management and first aid and environmental sanitation (water, toilet, food, garbage)

As per WHO, there are 10 core health issues that need to be addressed during emergencies. These are:

☆ *Assessment of health risks*: Analysis of health information, health risks and needs is essential for effective planning of the public health interventions needed to prevent or alleviate the impact of emergencies on physical and mental health.

☆ *Health co-ordination*: There should be co ordination between the health activities of governments, UN agencies and non-governmental organizations to ensure they are in line with international standards and local priorities and do not compromise or damage longer term health development.

☆ *Epidemic and nutritional surveillance*: The surveillance systems need to be strengthened and integrate information from external partners so that the
Figure 16.9arliest, conditions related to childbirth and damage to mental health. possible action can be taken against communicable diseases, common childhood illnesses, malnutrition

☆ *Control of preventable causes of illness and death*: Health priorities should be identified and addressed. These include prevention and response to infectious diseases from HIV/AIDS and tuberculosis to measles and other childhood diseases, mental health, environmental health, water, food, shelter, sanitation and the violence and injury prevention as well as all aspects of health care delivery. Safe water supply can be ensured by adequate chlorination, Water quality monitoring, Recording and reporting of water borne diseases and their management.

Vaccination against measles, typhoid, hepatitis and cholera is an effective strategy to prevent outbreaks of these diseases.

☆ *Access to basic preventative and curative care*: Ensure basic preventative and curative care is available, including access to good quality essential drugs and vaccines, surgical supplies and health information, to all affected and particularly to the especially vulnerable such as the very young, elderly, pregnant women, the disabled and the chronically ill.

☆ *Prevention of malnutrition*: Ensure that food distribution is equitable. Food safety and hygiene of food handlers are other issues. Special emphasis should be given on nutrition of infants and young children. Breast feeding should be encouraged in infants.

☆ *Management of health risks in the environment*: Environmental health risks in emergencies should be analysed and dealt with. Facilities should be provided for safe excreta disposal. Temporary latrines or trenches may have to be constructed. It has to be ensured that ground water is not contaminated. Cleanliness of these toilets can be ensured by involving people of the local community. The risk of vector borne diseases also increases during disasters. Strict vector control measures should be taken up. These include reduction of breeding sites by draining water collections, Insecticide sprays and providing mosquito nets and insect repellants to the people

☆ *Protection of health workers, services and structures*: WHO acts as an advocate for national and international health workers in situations of crisis, and is a key partner in negotiating secure humanitarian access and protecting the neutrality of health workers, services and structures.

☆ *Human rights to health*: Where basic human rights such as access to health, food, nutrition or education are unfulfilled, people are more vulnerable to the negative health impact of natural or man-made disasters. Ensure that humanitarian health activities combine the best public health practice with adherence to human rights principles regardless of adverse political or natural environments.

☆ *Reducing the impact of future crises*: Beyond the acute crisis, health authorities in disaster-prone and vulnerable areas should work to develop strategies for rapid response, building up an experienced cadre of national staff resilient to emergencies, and crucially taking action to reduce the impact of disaster before it strikes.[5,6]

Rehabilitation Phase

This phase aims at restoring the community to pre disaster levels. This phase starts right from the emergency phase and requires multi sectoral efforts. Coordinated inputs from departments like health, public works and engineering, transport, power, information and planning are required to reinstate the infrastructure and functioning of the affected community.

Health Facilities during Disaster

Health systems are also among the most vulnerable to natural disasters. After the 2004 Indian Ocean tsunami, a large number of health institutions were damaged. These included hospitals, drug stores, cold rooms, preventive health care offices, health staff accommodation facilities and district health offices. In addition, a large number of vehicles (ambulances, vans, motorbikes) and most of the medical equipment and office equipment in the affected areas were totally destroyed. The loss of health personnel included medical officers, nurses, midwives and support staff. Furthermore, a large number of health staff were injured, traumatized or displaced by the event, hence were unable to assist the affected.

On this World Health Day WHO is focusing attention on the large numbers of lives that can be saved during earthquakes, floods, conflicts and other emergencies through better design and construction of health facilities and by preparing and training health staff. World Health Day 2009 focuses on the safety of health facilities and the readiness of health workers who treat those affected by emergencies. Health centres and staff are critical lifelines for vulnerable people in disasters - treating injuries, preventing illnesses and caring for people's health needs. Often, already fragile health systems are unable to keep functioning through a disaster, with immediate and future public health consequences. The official slogan is "Save lives. Make hospitals safe in emergencies."[7]

A well documented and tested disaster management plan (DMP) to be in place in every hospital. To increase their preparedness for mass casualties, hospitals have to expand their focus to include both internal and community-level planning. The disaster management plan of a hospital should incorporate various issues that address natural disasters; biological, chemical, nuclear-radiological and explosive-incendiary terrorism incidents; collaboration with outside organizations for planning; establishment of alternate care sites; clinician training in the management of exposures to weaponizable infectious diseases, chemicals and nuclear materials; drills on aspects of the response plans; and equipment and bed capacity available at the hospital. The most important external agencies for collaboration would be state and local public health departments, emergency medical services, fire departments and law enforcing agencies. The key hospital personnel should be trained to implement a formal incident command system, which is an organized procedure for managing resources and personnel during an emergency. The hospitals should also have adequate availability of personal protective hazardous materials suits, negative pressure isolation rooms and decontamination showers. A hospital's emergency response plan has to be evaluated whether that plan addresses these issues[8].

Disaster Mitigation

Mitigation efforts attempt to prevent hazards from developing into disasters altogether, or to reduce the effects of disasters when they occur. The mitigation phase differs from the other phases because it focuses on long-term measures for reducing or eliminating risk. The implementation of mitigation strategies can be considered a part of the recovery process if applied after a disaster occurs. Mitigative measures can be structural or non-structural. Structural measures use technological solutions, like flood levees. Non-structural measures include legislation, land-use planning (*e.g.* the designation of nonessential land like parks to be used as flood zones), and insurance. Mitigation is the most cost-efficient method for reducing the impact of hazards. They also include vulnerability reduction measures such as awareness raising, improving community health security, and relocation or protection of vulnerable populations or structures. Mitigation measures on individual structures can be achieved by design standards, building codes and performance specifications. Building codes, critical front-line defence for achieving stronger engineered structures, need to be drawn up in accordance with the vulnerability of the area and implemented through appropriate techno-legal measures.[1,3]

Disaster Preparedness

Disaster preparedness is a program of long-term activities whose goals are to strengthen the overall capacity and capability of a country or a community to manage efficiently all types of emergencies and bring about an orderly transition from relief through recovery, and back to sustained development. It requires that emergency plans be developed, personnel at all levels and in all sectors be trained, and communities at risk be educated, and that these measures be monitored and evaluated regularly.

Disaster preparedness is an ongoing multi-sectoral activity.[1,3] It involves following activities:

- ☆ The focus towards preventive disaster management and development of a national ethos of prevention calls for an awareness generation at all levels.

- ☆ Capacity building should not be limited to professionals and personnel involved in disaster management but should also focus on building the knowledge, attitude and skills of a community to cope with the effects of disasters. Identification and training of volunteers from the community towards first response measures as well as mitigation measures is an urgent imperative.

- ☆ Within a vulnerable community, there exist groups that are more vulnerable like women and children, aged and infirm and physically challenged people who need special care and attention especially during disaster situations. Efforts are required for identifying such vulnerable groups and providing special assistance in terms of evacuation, relief, aid and medical attention to them in disaster situations.

Our country has the integrated administrative machinery for management of disasters supplements the efforts of the States by providing financial and logistic support.

Ministries Responsible for Various Categories of Disasters

The Ministry of Home Affairs is the nodal Ministry for coordination of relief and response and overall natural disaster management, and the Department of Agriculture and Cooperation is the nodal Ministry for drought management. Other Ministries are assigned the responsibility of providing emergency support in case of disasters that fall in their purview.

Disaster	Nodal Ministry
Natural Disasters Management (other than Drought)	Ministry of Home Affairs
Drought Relief	Ministry of Agriculture
Air Accidents	Ministry of Civil Aviation
Railway Accidents	Ministry of Railways
Chemical Disasters	Ministry of Environment and Forests
Biological Disasters	Ministry of Health
Nuclear Disasters	Department of Atomic Energy

Conclusion

Over the past 30 years, there has been a major shift in how emergencies and crises are managed. More emphasis used to be placed on humanitarian response and relief activities national or international – with little attention given to strategies and actions in place prior to disasters that can mitigate the effects of these events on communities and preserve lives and assets. It is becoming increasingly clear that while humanitarian efforts remain important and need continued attention, community-based risk reduction and emergency preparedness programs are critical for reducing the effects of emergencies, disasters and other crises, and thus essential for the attainment and protection of sustainable development[3]. Vulnerability of community and risk factors which can give rise to more health damage need to be identified and appropriate measures should be undertaken to remove such risks. So public health demands the shift in the disaster management from mitigation to risk reduction.

International initiatives by the humanitarian community are geared increasingly towards supporting this objective. The challenge is to put in place systematic capacities such as legislation, plans, coordination mechanisms and procedures, institutional capacities and budgets, skilled personnel, information, and public awareness and participation that can measurably reduce future risks and losses.

References

WHO. Risk reduction and emergency preparedness : WHO six-year strategy for the health sector and community capacity development.WHO.2007

Disaster Management: Expert guidelines on providing healthcare. Learning Objectives: After completing this module, you will learn: *www.who.int/ehealth/srilanka/documents/Disaster_Management_eLearning_Module_1.pdf* -

National institute of Disaster Management. Ministry of Home Affairs. Management of natural disasters. Available online at: www.nidm.net.in

Kishore J. A Dictionary of Public Health. New Delhi: Century Publications 2007.

WHO. Emergency and Humanitarian Action.2001. Available from: www.who.int

Kishore J. National emergency Preparedness Plan. In: National Health Programs of India. National Policies and Legislations Related to Health. Seventh Edition. Century Publications. 2008.

WHO. World Health Day 2009. Available from: www.who.int/world-health-day/

Mehta S. Disaster and mass casualty management in a hospital: How well are we prepared? J Postgrad Med [serial online] 2006 [cited 2009 Apr 19];52:89-90. Available from: http://www.jpgmonline.com/text.asp?2006/52/2/89/25148.

17

Role of Stakeholders in Disaster Responses

Avanish Kumar, Meena Raghunathan and Yogita Nandnwar

In India, the devastation that could be caused by industrial disaster became poignantly evident during the Bhopal Gas Tragedy in 1984. Following Bhopal, laws and regulations have been made more stringent, but the state of industrial safety and management measures on the ground are such that chances of major and minor accidents, in almost every corner of the country are considerable. And it would be communities near the industries which would probably bear the brunt of the impacts of such accidents. Community preparation is a significant, but often overlooked or underemphasized input for reducing disaster impacts.

It is in this background that a project was undertaken by Centre for Environment Education, India titled 'Testing Communication Strategies for Industrial Disaster Risk Reduction' supported by the ProVention Consortium. The project basically looked at evolving a model for community disaster preparedness. The model had the school playing a pivotal role in the communication strategy, and involved several other stakeholders.

The project focused on industrial disaster preparedness in one of the industrial estates of Gujarat state of India. Gujarat is one of the most industrialized states of the country, with close to 180 industrial estates. Each estate has hundreds of small to medium scale industries. There are several settlements very close to the Estates. These are mostly occupied by workers employed in the industries. Their children often go to schools which are also situated close to the Estate. The project focused on these schools as channels for information about preparedness among themselves and to reach out to the community. The project was quite successful in meeting its objectives and reached out through various capacity-building and awareness programmes using different medium and approaches like a teacher's manual, posters, skit, kites, etc.,. The message was effectively communicated through the schools to a larger fraternity in the estate.

Apart from the schools, several other stakeholders related to the issue, including the Industries Association, government departments like the Factory Inspectorate, Gujarat Industrial Development Corporation, the Fire Brigade department and also local NGOs and other institutions working in this field were involved. For each one of these, a different approach was adopted to gain their involvement and ownership in the project.

Schools were the first one to be approached. Six schools were selected for the project, out of which five were 'private' schools. While we were initially skeptical about how the schools would receive this idea, as it involved giving up certain hours of teaching time, and called for extra initiative on the part of the teachers to do the project activities. We were quite surprised to see the enthusiasm with which all the schools responded to the idea. Several Principals and teachers said that something like this was urgently needed, and this was a very timely activity. Starting with an orientation session to initiate the process, the schools were involved in the project activities over a six month period. CEE's experience of working with schools for more than twenty years was what probably helped us in gaining a favourable ground with the schools very early in the project.

Another important stakeholder in the project was the Industries Association. When the Association was first approached for the project, they were somewhat skeptical. The Association office-bearers were very cooperative but had the apprehension whether the project would strengthen the perception that industries are polluting and dangerous. We were able to convince them that our project would do help to dispel such doubts and would result in better preparation among the community and the workers for any accident or disaster in the industry and the Estate. With these discussions, the Association was cooperative and gave support to the project activities actively. We also agreed that drafts of all publications or any other material that we would bring out for the project would be shown to them prior to publication. We acknowledged their cooperation in all activities and kept them informed about the progress of the project. This helped in enhancing their role and building a sense of ownership among them for the project.

The other important stakeholders were the government agencies. Even as we started visiting the Estate and were identifying the schools, we visited the GIDC office and the Factory Inspectorate. We sought their guidance for implementing the project. Again, at every stage of the project, we kept them informed and they provided useful support towards the activities. They commented on the draft manual and provided useful insights. They were also helpful for giving the information about the Local Crisis Group for the Vatva Industrial Estate.

The Ahmedabad Fire Brigade was another important stakeholder. The approach followed here was also to involve them as an inherent part of the project and not as an external agency doing activities for the project. As awareness raising and capacity-building the citizens is part of their mandate, we could find common ground and they come to see the project as helping them in furthering their objectives. We received encouraging cooperation, and for many activities they went out of their way to accommodate us. One of the highlights of a concluding event we did at the Estate was a demonstration by the Fire Brigade personnel about various fire-fighting and industrial disaster fighting drills. More than 50 personnel with all their gadgets and equipments were present for this demonstration, and they did not charge us anything for carrying out this huge exercise.

One way in which we involved these agencies was to make key persons judges at the final event for the different competitions that were held for the students as a part of the concluding event. This event was also the forum where these agencies shared their experiences from the project and looked at how it can be sustained and scaled-up. A key to working with partners was not to confront them but try to find common grounds of concern and synergy.

All these approaches helped in providing sustainability to the project activities. The ownership of the project has, over the course of the project slowly shifted to one of the most important players in the project–the schools. Through the orientation sessions, the teacher's workshop and finally the awareness programmes by students in schools and further to community, the school management, teachers and the students have really begun to take ownership of activities of the project. The programme sustainability would ensured as the manual and other information developed would be available with the schools, and they will have trained teachers to sustain the effort. Hence, every new batch of students would get trained in this preparedness activity, and further, spreading it to the community. This will also make the material developed as part of the project very cost-effective as it will be repeatedly utilized. The teachers and students propagating this already feel a sense of ownership in the whole exercise and would act as leaders in case of an event.

Other key institutions, as detailed earlier, have also started to develop a sense of ownership for the project. The Association, by the end of the project started feeling that this activity is beneficial for the welfare of their Estate and can bring credit to them for being a safety-friendly industrial estate. An example to illustrate this ownership and appreciation for the project activities came through when the Association offered to sponsor the refreshments for more than 300 participants in the concluding event of the project.

Overall the important learning has been that participation from all relevant stakeholders not only helps in carrying out the activities during the project period, but is also crucial for the long-term sustainability.